# Human Impacts on Our Climate

Grade 6

## STEM Road Map for Middle School

Edited by Carla C. Johnson, Janet B. Walton, and Erin Peters-Burton

**nsta** Press
National Science Teaching Association
Arlington, Virginia

Claire Reinburg, Director
Rachel Ledbetter, Managing Editor
Jennifer Merrill, Associate Editor
Andrea Silen, Associate Editor
Donna Yudkin, Book Acquisitions Manager

ART AND DESIGN
Will Thomas Jr., Director, cover
Himabindu Bichali, Graphic Designer, interior
  design

PRINTING AND PRODUCTION
Catherine Lorrain, Director

NATIONAL SCIENCE TEACHING ASSOCIATION
1840 Wilson Blvd., Arlington, VA 22201
*www.nsta.org/store*
For customer service inquiries, please call 800-277-5300.

*NSTA is committed to publishing material that promotes the best in inquiry-based science education. However, conditions of actual use may vary, and the safety procedures and practices described in this book are intended to serve only as a guide. Additional precautionary measures may be required. NSTA and the authors do not warrant or represent that the procedures and practices in this book meet any safety code or standard of federal, state, or local regulations. NSTA and the authors disclaim any liability for personal injury or damage to property arising out of or relating to the use of this book, including any of the recommendations, instructions, or materials contained therein.*

**Library of Congress Cataloging-in-Publication Data**
Names: Johnson, Carla C., 1969- editor. | Walton, Janet B., 1968- editor. | Peters-Burton, Erin E., editor.
Title: Human impacts on our climate, grade 6 : STEM road map for middle school / edited by Carla C. Johnson, Janet B. Walton, and Erin Peters-Burton.
Description: Arlington, VA : National Science Teaching Association, [2020] | Includes bibliographical references and index.
Identifiers: LCCN 2019056397 (print) | LCCN 2019056398 (ebook) | ISBN 9781681404080 (paperback) | ISBN 9781681404097 (pdf)
Subjects: LCSH: Climatic changes--Effect of human beings on. | Climatic changes--Study and teaching (Middle school)
Classification: LCC QC903 .I1855 2020 (print) | LCC QC903 (ebook) | DDC 363.738/740712--dc23
LC record available at *https://lccn.loc.gov/2019056397*
LC ebook record available at *https://lccn.loc.gov/2019056398*

# Human Impacts on Our Climate

STEM Road Map
for Middle School

Grade
6

# CONTENTS

## Part 1: The STEM Road Map: Background, Theory, and Practice

## Part 2: Human Impacts on Our Climate: STEM Road Map Module

# CONTENTS

# ABOUT THE EDITORS AND AUTHORS

**Dr. Carla C. Johnson** is a professor of science education in the College of Education and Office of Research and Innovation Faculty Research Fellow at North Carolina State University in Raleigh. She was most recently an associate dean, provost fellow, and professor of science education at Purdue University in West Lafayette, Indiana. Dr. Johnson serves as the director of research and evaluation for the Department of Defense–funded Army Educational Outreach Program (AEOP), a global portfolio of STEM education programs, competitions, and apprenticeships. She has been a leader in STEM education for the past decade, serving as the director of STEM Centers, editor of the *School Science and Mathematics* journal, and lead researcher for the evaluation of Tennessee's Race to the Top–funded STEM portfolio. Dr. Johnson has published over 100 articles, books, book chapters, and curriculum books focused on STEM education. She is a former science and social studies teacher and was the recipient of the 2013 Outstanding Science Teacher Educator of the Year award from the Association for Science Teacher Education (ASTE), the 2012 Award for Excellence in Integrating Science and Mathematics from the School Science and Mathematics Association (SSMA), the 2014 award for best paper on Implications of Research for Educational Practice from ASTE, and the 2006 Outstanding Early Career Scholar Award from SSMA. Her research focuses on STEM education policy implementation, effective science teaching, and integrated STEM approaches.

**Dr. Janet B. Walton** is a senior research scholar and the assistant director of evaluation for AEOP in the College of Education at North Carolina State University. She merges her economic development and education backgrounds to develop K–12 curricular materials that integrate real-life issues with sound cross-curricular content. Her research focuses on mixed methods research methodologies and collaboration between schools and community stakeholders for STEM education and problem- and project-based learning pedagogies. With this research agenda, she works to bring contextual STEM experiences into the classroom and provide students and educators with innovative resources and curricular materials.

**Dr. Erin Peters-Burton** is the Donna R. and David E. Sterling Endowed Professor in Science Education in the College of Education and Human Development at George Mason University in Fairfax, Virginia. She uses her experiences from 15 years as an engineer and secondary science, engineering, and mathematics teacher to develop research projects that directly inform classroom practice in science and engineering. Her research agenda is based on the idea that all students should build self-awareness of how they learn science and engineering. She works to help students see themselves as "science-minded" and help teachers create classrooms that support student skills to develop scientific knowledge. To accomplish this, she pursues research projects that investigate ways that students and teachers can use self-regulated learning theory in science and engineering, as well as how inclusive STEM schools can help students succeed. During her tenure as a secondary teacher, she held a National Board Certification in Early Adolescent Science and was an Albert Einstein Distinguished Educator Fellow for NASA. As a researcher, Dr. Peters-Burton has published over 100 articles, books, book chapters, and curriculum books focused on STEM education and educational psychology. She received the Outstanding Science Teacher Educator of the Year award from ASTE and a Teacher of Distinction Award and a Scholarly Achievement Award from George Mason University in 2012, and in 2010 she was named University Science Educator of the Year by the Virginia Association of Science Teachers.

**Dr. Toni Ivey** is an associate professor of science education in the College of Education, Health and Aviation at Oklahoma State University. Dr. Ivey is a former science teacher whose research is focused on science and STEM education for students and teachers across K–20.

**Dr. Tamara J. Moore** is an associate professor of engineering education in the College of Engineering at Purdue University. Dr. Moore's research focuses on defining STEM integration through the use of engineering as the connection and investigating its power for student learning.

**Dr. Sue Christian Parsons** is an associate professor and the Jacques Munroe Professor in Reading and Literacy Education at Oklahoma State University. A former English language arts teacher, she concentrates her research on teacher development and teaching and advocating for diverse learners through literature for children and young adults.

**Dr. Adrienne Redmond-Sanogo** is an associate dean for academic affairs and associate professor of mathematics education in the College of Education at Oklahoma State University. Dr. Redmond-Sanogo's research focuses on mathematics and STEM education across K–12 and preservice teacher education.

**Dr. Toni A. Sondergeld** is an associate professor of assessment, research, and statistics in the School of Education at Drexel University in Philadelphia. A former elementary and middle school science and mathematics teacher, Dr. Sondergeld centers her research on assessment and evaluation in education, with a focus on K–12 STEM.

**Dr. Juliana Utley** is a professor and Morsani Chair in Mathematics Education in the College of Education, Health and Aviation at Oklahoma State University. A former mathematics teacher, Dr. Utley focuses her research on mathematics and STEM education across K–12.

# ACKNOWLEDGMENTS

This module was developed as a part of the STEM Road Map project (Carla C. Johnson, principal investigator). The Purdue University College of Education, General Motors, and other sources provided funding for this project.

Copyright © 2015 from *STEM Road Map: A Framework for Integrated STEM Education*, edited by C. C. Johnson, E. E. Peters-Burton, and T. J. Moore. Reproduced by permission of Taylor and Francis Group, LLC, a division of Informa plc.

See *www.routledge.com/products/9781138804234* for more information about *STEM Road Map: A Framework for Integrated STEM Education*.

# PART 1

# THE STEM ROAD MAP

## BACKGROUND, THEORY, AND PRACTICE

# OVERVIEW OF THE *STEM ROAD MAP CURRICULUM SERIES*

*Carla C. Johnson, Erin Peters-Burton, and Tamara J. Moore*

The *STEM Road Map Curriculum Series* was conceptualized and developed by a team of STEM educators from across the United States in response to a growing need to infuse real-world learning contexts, delivered through authentic problem-solving pedagogy, into K–12 classrooms. The curriculum series is grounded in integrated STEM, which focuses on the integration of the STEM disciplines—science, technology, engineering, and mathematics—delivered across content areas, incorporating the Framework for 21st Century Learning along with grade-level-appropriate academic standards.

The curriculum series begins in kindergarten, with a five-week instructional sequence that introduces students to the STEM themes and gives them grade-level-appropriate topics and real-world challenges or problems to solve. The series uses project-based and problem-based learning, presenting students with the problem or challenge during the first lesson, and then teaching them science, social studies, English language arts, mathematics, and other content, as they apply what they learn to the challenge or problem at hand.

Authentic assessment and differentiation are embedded throughout the modules. Each *STEM Road Map Curriculum Series* module has a lead discipline, which may be science, social studies, English language arts, or mathematics. All disciplines are integrated into each module, along with ties to engineering. Another key component is the use of STEM Research Notebooks to allow students to track their own learning progress. The modules are designed with a scaffolded approach, with increasingly complex concepts and skills introduced as students progress through grade levels.

The developers of this work view the curriculum as a resource that is intended to be used either as a whole or in part to meet the needs of districts, schools, and teachers who are implementing an integrated STEM approach. A variety of implementation formats are possible, from using one stand-alone module at a given grade level to using all five modules to provide 25 weeks of instruction. Also, within each grade band (K–2, 3–5, 6–8, 9–12), the modules can be sequenced in various ways to suit specific needs.

## STANDARDS-BASED APPROACH

The *STEM Road Map Curriculum Series* is anchored in the *Next Generation Science Standards (NGSS)*, the *Common Core State Standards for Mathematics (CCSS Mathematics)*, the *Common Core State Standards for English Language Arts (CCSS ELA)*, and the Framework for 21st Century Learning. Each module includes a detailed curriculum map that incorporates the associated standards from the particular area correlated to lesson plans. The STEM Road Map has very clear and strong connections to these academic standards, and each of the grade-level topics was derived from the mapping of the standards to ensure alignment among topics, challenges or problems, and the required academic standards for students. Therefore, the curriculum series takes a standards-based approach and is designed to provide authentic contexts for application of required knowledge and skills.

## THEMES IN THE *STEM ROAD MAP CURRICULUM SERIES*

The K–12 STEM Road Map is organized around five real-world STEM themes that were generated through an examination of the big ideas and challenges for society included in STEM standards and those that are persistent dilemmas for current and future generations:

- Cause and Effect
- Innovation and Progress
- The Represented World
- Sustainable Systems
- Optimizing the Human Experience

These themes are designed as springboards for launching students into an exploration of real-world learning situated within big ideas. Most important, the five STEM Road Map themes serve as a framework for scaffolding STEM learning across the K–12 continuum.

The themes are distributed across the STEM disciplines so that they represent the big ideas in science (Cause and Effect; Sustainable Systems), technology (Innovation and Progress; Optimizing the Human Experience), engineering (Innovation and Progress; Sustainable Systems; Optimizing the Human Experience), and mathematics (The Represented World), as well as concepts and challenges in social studies and 21st century skills that are also excellent contexts for learning in English language arts. The process of developing themes began with the clustering of the *NGSS* performance expectations and the National Academy of Engineering's grand challenges for engineering, which led to the development of the challenge in each module and connections of the module activities to the *CCSS Mathematics* and *CCSS ELA* standards. We performed these

mapping processes with large teams of experts and found that these five themes provided breadth, depth, and coherence to frame a high-quality STEM learning experience from kindergarten through 12th grade.

## Cause and Effect

The concept of cause and effect is a powerful and pervasive notion in the STEM fields. It is the foundation of understanding how and why things happen as they do. Humans spend considerable effort and resources trying to understand the causes and effects of natural and designed phenomena to gain better control over events and the environment and to be prepared to react appropriately. Equipped with the knowledge of a specific cause-and-effect relationship, we can lead better lives or contribute to the community by altering the cause, leading to a different effect. For example, if a person recognizes that irresponsible energy consumption leads to global climate change, that person can act to remedy his or her contribution to the situation. Although cause and effect is a core idea in the STEM fields, it can actually be difficult to determine. Students should be capable of understanding not only when evidence points to cause and effect but also when evidence points to relationships but not direct causality. The major goal of education is to foster students to be empowered, analytic thinkers, capable of thinking through complex processes to make important decisions. Understanding causality, as well as when it cannot be determined, will help students become better consumers, global citizens, and community members.

## Innovation and Progress

One of the most important factors in determining whether humans will have a positive future is innovation. Innovation is the driving force behind progress, which helps create possibilities that did not exist before. Innovation and progress are creative entities, but in the STEM fields, they are anchored by evidence and logic, and they use established concepts to move the STEM fields forward. In creating something new, students must consider what is already known in the STEM fields and apply this knowledge appropriately. When we innovate, we create value that was not there previously and create new conditions and possibilities for even more innovations. Students should consider how their innovations might affect progress and use their STEM thinking to change current human burdens to benefits. For example, if we develop more efficient cars that use by-products from another manufacturing industry, such as food processing, then we have used waste productively and reduced the need for the waste to be hauled away, an indirect benefit of the innovation.

## The Represented World

When we communicate about the world we live in, how the world works, and how we can meet the needs of humans, sometimes we can use the actual phenomena to explain a concept. Sometimes, however, the concept is too big, too slow, too small, too fast, or too complex for us to explain using the actual phenomena, and we must use a representation or a model to help communicate the important features. We need representations and models such as graphs, tables, mathematical expressions, and diagrams because it makes our thinking visible. For example, when examining geologic time, we cannot actually observe the passage of such large chunks of time, so we create a timeline or a model that uses a proportional scale to visually illustrate how much time has passed for different eras. Another example may be something too complex for students at a particular grade level, such as explaining the $p$ subshell orbitals of electrons to fifth graders. Instead, we use the Bohr model, which more closely represents the orbiting of planets and is accessible to fifth graders.

When we create models, they are helpful because they point out the most important features of a phenomenon. We also create representations of the world with mathematical functions, which help us change parameters to suit the situation. Creating representations of a phenomenon engages students because they are able to identify the important features of that phenomenon and communicate them directly. But because models are estimates of a phenomenon, they leave out some of the details, so it is important for students to evaluate their usefulness as well as their shortcomings.

## Sustainable Systems

From an engineering perspective, the term *system* refers to the use of "concepts of component need, component interaction, systems interaction, and feedback. The interaction of subcomponents to produce a functional system is a common lens used by all engineering disciplines for understanding, analysis, and design" (Koehler, Bloom, and Binns 2013, p. 8). Systems can be either open (e.g., an ecosystem) or closed (e.g., a car battery). Ideally, a system should be sustainable, able to maintain equilibrium without much energy from outside the structure. Looking at a garden, we see flowers blooming, weeds sprouting, insects buzzing, and various forms of life living within its boundaries. This is an example of an ecosystem, a collection of living organisms that survive together, functioning as a system. The interaction of the organisms within the system and the influences of the environment (e.g., water, sunlight) can maintain the system for a period of time, thus demonstrating its ability to endure. Sustainability is a desirable feature of a system because it allows for existence of the entity in the long term.

In the STEM Road Map project, we identified different standards that we consider to be oriented toward systems that students should know and understand in the K–12 setting. These include ecosystems, the rock cycle, Earth processes (such as erosion, tectonics, ocean currents, weather phenomena), Earth-Sun-Moon cycles, heat transfer, and the interaction among the geosphere, biosphere, hydrosphere, and atmosphere. Students and teachers should understand that we live in a world of systems that are not independent of each other, but rather are intrinsically linked such that a disruption in one part of a system will have reverberating effects on other parts of the system.

## Optimizing the Human Experience

Science, technology, engineering, and mathematics as disciplines have the capacity to continuously improve the ways humans live, interact, and find meaning in the world, thus working to optimize the human experience. This idea has two components: being more suited to our environment and being more fully human. For example, the progression of STEM ideas can help humans create solutions to complex problems, such as improving ways to access water sources, designing energy sources with minimal impact on our environment, developing new ways of communication and expression, and building efficient shelters. STEM ideas can also provide access to the secrets and wonders of nature. Learning in STEM requires students to think logically and systematically, which is a way of knowing the world that is markedly different from knowing the world as an artist. When students can employ various ways of knowing and understand when it is appropriate to use a different way of knowing or integrate ways of knowing, they are fully experiencing the best of what it is to be human. The problem-based learning scenarios provided in the STEM Road Map help students develop ways of thinking like STEM professionals as they ask questions and design solutions. They learn to optimize the human experience by innovating improvements in the designed world in which they live.

## THE NEED FOR AN INTEGRATED STEM APPROACH

At a basic level, STEM stands for science, technology, engineering, and mathematics. Over the past decade, however, STEM has evolved to have a much broader scope and broader implications. Now, educators and policy makers refer to STEM as not only a concentrated area for investing in the future of the United States and other nations but also as a domain and mechanism for educational reform.

The good intentions of the recent decade-plus of focus on accountability and increased testing has resulted in significant decreases not only in instructional time for teaching science and social studies but also in the flexibility of teachers to promote authentic, problem solving–focused classroom environments. The shift has had a detrimental impact on student acquisition of vitally important skills, which many refer to as 21st century skills, and often the ability of students to "think." Further, schooling has become increasingly siloed into compartments of mathematics, science, English language arts, and social studies, lacking any of the connections that are overwhelmingly present in

the real world around children. Students have experienced school as content provided in boxes that must be memorized, devoid of any real-world context, and often have little understanding of why they are learning these things.

STEM-focused projects, curriculum, activities, and schools have emerged as a means to address these challenges. However, most of these efforts have continued to focus on the individual STEM disciplines (predominantly science and engineering) through more STEM classes and after-school programs in a "STEM enhanced" approach (Breiner et al. 2012). But in traditional and STEM enhanced approaches, there is little to no focus on other disciplines that are integral to the context of STEM in the real world. Integrated STEM education, on the other hand, infuses the learning of important STEM content and concepts with a much-needed emphasis on 21st century skills and a problem- and project-based pedagogy that more closely mirrors the real-world setting for society's challenges. It incorporates social studies, English language arts, and the arts as pivotal and necessary (Johnson 2013; Rennie, Venville, and Wallace 2012; Roehrig et al. 2012).

## FRAMEWORK FOR STEM INTEGRATION IN THE CLASSROOM

The *STEM Road Map Curriculum Series* is grounded in the Framework for STEM Integration in the Classroom as conceptualized by Moore, Guzey, and Brown (2014) and Moore et al. (2014). The framework has six elements, described in the context of how they are used in the *STEM Road Map Curriculum Series* as follows:

1. The STEM Road Map contexts are meaningful to students and provide motivation to engage with the content. Together, these allow students to have different ways to enter into the challenge.

2. The STEM Road Map modules include engineering design that allows students to design technologies (i.e., products that are part of the designed world) for a compelling purpose.

3. The STEM Road Map modules provide students with the opportunities to learn from failure and redesign based on the lessons learned.

4. The STEM Road Map modules include standards-based disciplinary content as the learning objectives.

5. The STEM Road Map modules include student-centered pedagogies that allow students to grapple with the content, tie their ideas to the context, and learn to think for themselves as they deepen their conceptual knowledge.

6. The STEM Road Map modules emphasize 21st century skills and, in particular, highlight communication and teamwork.

All of the STEM Road Map modules incorporate these six elements; however, the level of emphasis on each of these elements varies based on the challenge or problem in each module.

## THE NEED FOR THE *STEM ROAD MAP CURRICULUM SERIES*

As focus is increasing on integrated STEM, and additional schools and programs decide to move their curriculum and instruction in this direction, there is a need for high-quality, research-based curriculum designed with integrated STEM at the core. Several good resources are available to help teachers infuse engineering or more STEM enhanced approaches, but no curriculum exists that spans K–12 with an integrated STEM focus. The next chapter provides detailed information about the specific pedagogy, instructional strategies, and learning theory on which the *STEM Road Map Curriculum Series* is grounded.

## REFERENCES

Breiner, J., M. Harkness, C. C. Johnson, and C. Koehler. 2012. What is STEM? A discussion about conceptions of STEM in education and partnerships. *School Science and Mathematics* 112 (1): 3–11.

Johnson, C. C. 2013. Conceptualizing integrated STEM education: Editorial. *School Science and Mathematics* 113 (8): 367–368.

Koehler, C. M., M. A. Bloom, and I. C. Binns. 2013. Lights, camera, action: Developing a methodology to document mainstream films' portrayal of nature of science and scientific inquiry. *Electronic Journal of Science Education* 17 (2).

Moore, T. J., S. S. Guzey, and A. Brown. 2014. Greenhouse design to increase habitable land: An engineering unit. *Science Scope* 37 (7): 51–57.

Moore, T. J., M. S. Stohlmann, H. H. Wang, K. M. Tank, A. W. Glancy, and G. H. Roehrig. 2014. Implementation and integration of engineering in K–12 STEM education. In *Engineering in pre-college settings: Synthesizing research, policy, and practices*, ed. S. Purzer, J. Strobel, and M. Cardella, 35–60. West Lafayette, IN: Purdue Press.

Rennie, L., G. Venville, and J. Wallace. 2012. *Integrating science, technology, engineering, and mathematics: Issues, reflections, and ways forward.* New York: Routledge.

Roehrig, G. H., T. J. Moore, H. H. Wang, and M. S. Park. 2012. Is adding the E enough? Investigating the impact of K–12 engineering standards on the implementation of STEM integration. *School Science and Mathematics* 112 (1): 31–44.

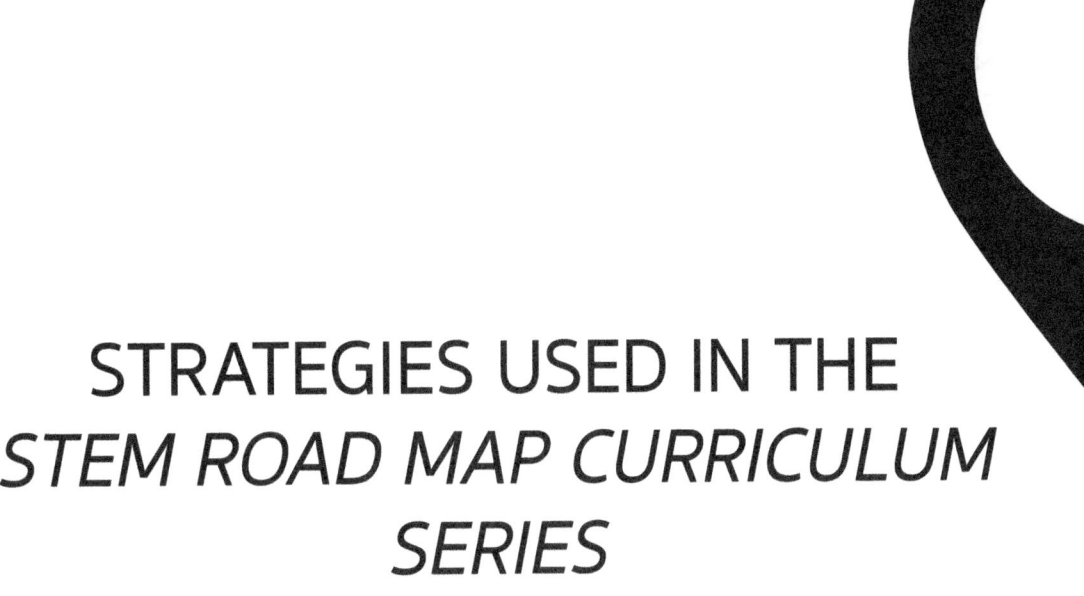

# STRATEGIES USED IN THE *STEM ROAD MAP CURRICULUM SERIES*

Erin Peters-Burton, Carla C. Johnson, Toni A. Sondergeld, and Tamara J. Moore

The *STEM Road Map Curriculum Series* uses what has been identified through research as best-practice pedagogy, including embedded formative assessment strategies throughout each module. This chapter briefly describes the key strategies that are employed in the series.

## PROJECT- AND PROBLEM-BASED LEARNING

Each module in the *STEM Road Map Curriculum Series* uses either project-based learning or problem-based learning to drive the instruction. Project-based learning begins with a driving question to guide student teams in addressing a contextualized local or community problem or issue. The outcome of project-based instruction is a product that is conceptualized, designed, and tested through a series of scaffolded learning experiences (Blumenfeld et al. 1991; Krajcik and Blumenfeld 2006). Problem-based learning is often grounded in a fictitious scenario, challenge, or problem (Barell 2006; Lambros 2004). On the first day of instruction within the unit, student teams are provided with the context of the problem. Teams work through a series of activities and use open-ended research to develop their potential solution to the problem or challenge, which need not be a tangible product (Johnson 2003).

## ENGINEERING DESIGN PROCESS

The *STEM Road Map Curriculum Series* uses engineering design as a way to facilitate integrated STEM within the modules. The engineering design process (EDP) used in the STEM Road Map series is depicted in Figure 2.1 (p. 10). It highlights two major aspects of engineering design—problem scoping and solution generation—and six specific components of working toward a design: define the problem, learn about the problem, plan a solution, try the solution, test the solution, decide whether the solution is good enough. It

**Figure 2.1. Engineering Design Process**

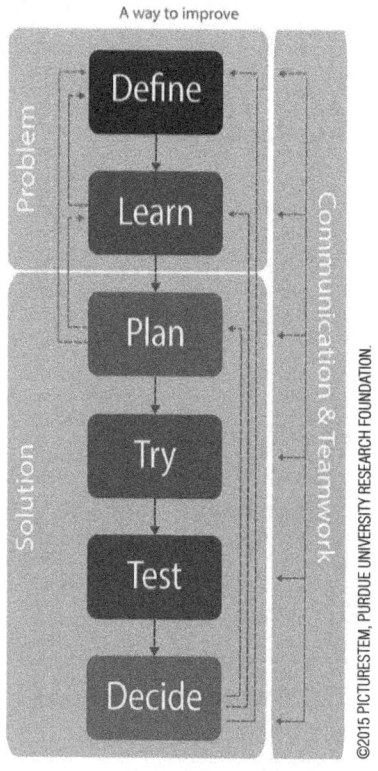

©2015 PICTURESTEM, PURDUE UNIVERSITY RESEARCH FOUNDATION.

also shows that communication and teamwork are involved throughout the entire process. As the arrows in the figure indicate, the order in which the components of engineering design are addressed depends on what becomes needed as designers progress through this EDP. Designers must communicate and work in teams throughout the process. An EDP is iterative, meaning that components of the process can be repeated as needed until the design is good enough to present to the client as a potential solution to the problem.

Problem scoping is the process of gathering and analyzing information to deeply understand the engineering design problem. It includes defining the problem and learning about the problem. Defining the problem includes identifying the problem, the client, and the end user of the design. The client is the person (or people) who hired the designers to do the work, and the end user is the person (or people) who will use the final design. The designers must also identify the criteria and the constraints of the problem. The criteria are the things the client wants from the solution, and the constraints are the things that limit the possible solutions. The designers must spend significant time learning about the problem, which can include activities such as the following:

- Reading informational texts and researching about relevant concepts or contexts

- Identifying and learning about needed mathematical and scientific skills, knowledge, and tools

- Learning about things done previously to solve similar problems

- Experimenting with possible materials that could be used in the design

Problem scoping also allows designers to consider how to measure the success of the design in addressing specific criteria and staying within the constraints over multiple iterations of solution generation.

Solution generation includes planning a solution, trying the solution, testing the solution, and deciding whether the solution is good enough. Planning the solution includes generating many design ideas that both address the criteria and meet the constraints. Here the designers must consider what was learned about the problem during problem scoping. Design plans include clear communication of design ideas through media such as notebooks, blueprints, schematics, or storyboards. They also include details about the design, such as measurements, materials, colors, costs of materials, instructions for how things fit together, and sets of directions. Making the decision about which design idea to move forward involves considering the trade-offs of each design idea.

Once a clear design plan is in place, the designers must try the solution. Trying the solution includes developing a prototype (a testable model) based on the plan generated. The prototype might be something physical or a process to accomplish a goal. This component of design requires that the designers consider the risk involved in implementing the design. The prototype developed must be tested. Testing the solution includes conducting fair tests that verify whether the plan is a solution that is good enough to meet the client and end user needs and wants. Data need to be collected about the results of the tests of the prototype, and these data should be used to make evidence-based decisions regarding the design choices made in the plan. Here, the designers must again consider the criteria and constraints for the problem.

Using the data gathered from the testing, the designers must decide whether the solution is good enough to meet the client and end user needs and wants by assessment based on the criteria and constraints. Here, the designers must justify or reject design decisions based on the background research gathered while learning about the problem and on the evidence gathered during the testing of the solution. The designers must now decide whether to present the current solution to the client as a possibility or to do more iterations of design on the solution. If they decide that improvements need to be made to the solution, the designers must decide if there is more that needs to be understood about the problem, client, or end user; if another design idea should be tried; or if more planning needs to be conducted on the same design. One way or another, more work needs to be done.

Throughout the process of designing a solution to meet a client's needs and wants, designers work in teams and must communicate to each other, the client, and likely the end user. Teamwork is important in engineering design because multiple perspectives and differing skills and knowledge are valuable when working to solve problems. Communication is key to the success of the designed solution. Designers must communicate their ideas clearly using many different representations, such as text in an engineering notebook, diagrams, flowcharts, technical briefs, or memos to the client.

## LEARNING CYCLE

The same format for the learning cycle is used in all grade levels throughout the STEM Road Map, so that students engage in a variety of activities to learn about phenomena in the modules thoroughly and have consistent experiences in the problem- and project-based learning modules. Expectations for learning by younger students are not as high as for older students, but the format of the progression of learning is the same. Students who have learned with curriculum from the STEM Road Map in early grades know

what to expect in later grades. The learning cycle consists of five parts—Introductory Activity/Engagement, Activity/Exploration, Explanation, Elaboration/Application of Knowledge, and Evaluation/Assessment—and is based on the empirically tested 5E model from BSCS (Bybee et al. 2006).

In the Introductory Activity/Engagement phase, teachers introduce the module challenge and use a unique approach designed to pique students' curiosity. This phase gets students to start thinking about what they already know about the topic and begin wondering about key ideas. The Introductory Activity/Engagement phase positions students to be confident about what they are about to learn, because they have prior knowledge, and clues them into what they don't yet know.

In the Activity/Exploration phase, the teacher sets up activities in which students experience a deeper look at the topics that were introduced earlier. Students engage in the activities and generate new questions or consider possibilities using preliminary investigations. Students work independently, in small groups, and in whole-group settings to conduct investigations, resulting in common experiences about the topic and skills involved in the real-world activities. Teachers can assess students' development of concepts and skills based on the common experiences during this phase.

During the Explanation phase, teachers direct students' attention to concepts they need to understand and skills they need to possess to accomplish the challenge. Students participate in activities to demonstrate their knowledge and skills to this point, and teachers can pinpoint gaps in student knowledge during this phase.

In the Elaboration/Application of Knowledge phase, teachers present students with activities that engage the students in higher-order thinking to create depth and breadth of student knowledge, while connecting ideas across topics within and across STEM. Students apply what they have learned thus far in the module to a new context or elaborate on what they have learned about the topic to a deeper level of detail.

In the last phase, Evaluation/Assessment, teachers give students summative feedback on their knowledge and skills as demonstrated through the challenge. This is not the only point of assessment (as discussed in the section on Embedded Formative Assessments), but it is an assessment of the culmination of the knowledge and skills for the module. Students demonstrate their cognitive growth at this point and reflect on how far they have come since the beginning of the module. The challenges are designed to be multidimensional in the ways students must collaborate and communicate their new knowledge.

## STEM RESEARCH NOTEBOOK

One of the main components of the *STEM Road Map Curriculum Series* is the STEM Research Notebook, a place for students to capture their ideas, questions, observations, reflections, evidence of progress, and other items associated with their daily work. At the beginning of each module, the teacher walks students through the setup of the STEM

Research Notebook, which could be a three-ring binder, composition book, or spiral notebook. You may wish to have students create divided sections so that they can easily access work from various disciplines during the module. Electronic notebooks kept on student devices are also acceptable and encouraged. Students will develop their own table of contents and create chapters in the notebook for each module.

Each lesson in the *STEM Road Map Curriculum Series* includes one or more prompts that are designed for inclusion in the STEM Research Notebook and appear as questions or statements that the teacher assigns to students. These prompts require students to apply what they have learned across the lesson to solve the big problem or challenge for that module. Each lesson is designed to meaningfully refer students to the larger problem or challenge they have been assigned to solve with their teams. The STEM Research Notebook is designed to be a key formative assessment tool, as students' daily entries provide evidence of what they are learning. The notebook can be used as a mechanism for dialogue between the teacher and students, as well as for peer and self-evaluation.

The use of the STEM Research Notebook is designed to scaffold student notebooking skills across the grade bands in the *STEM Road Map Curriculum Series*. In the early grades, children learn how to organize their daily work in the notebook as a way to collect their products for future reference. In elementary school, students structure their notebooks to integrate background research along with their daily work and lesson prompts. In the upper grades (middle and high school), students expand their use of research and data gathering through team discussions to more closely mirror the work of STEM experts in the real world.

## THE ROLE OF ASSESSMENT IN THE *STEM ROAD MAP CURRICULUM SERIES*

Starting in the middle years and continuing into secondary education, the word *assessment* typically brings grades to mind. These grades may take the form of a letter or a percentage, but they typically are used as a representation of a student's content mastery. If well thought out and implemented, however, classroom assessment can offer teachers, parents, and students valuable information about student learning and misconceptions that does not necessarily come in the form of a grade (Popham 2013).

The *STEM Road Map Curriculum Series* provides a set of assessments for each module. Teachers are encouraged to use assessment information for more than just assigning grades to students. Instead, assessments of activities requiring students to actively engage in their learning, such as student journaling in STEM Research Notebooks, collaborative presentations, and constructing graphic organizers, should be used to move student learning forward. Whereas other curriculum with assessments may include objective-type (multiple-choice or matching) tests, quizzes, or worksheets, we have intentionally avoided these forms of assessments to better align assessment strategies with teacher instruction and

student learning techniques. Since the focus of this book is on project- or problem-based STEM curriculum and instruction that focuses on higher-level thinking skills, appropriate and authentic performance assessments were developed to elicit the most reliable and valid indication of growth in student abilities (Brookhart and Nitko 2008).

## Comprehensive Assessment System

Assessment throughout all STEM Road Map curriculum modules acts as a comprehensive system in which formative and summative assessments work together to provide teachers with high-quality information on student learning. Formative assessment occurs when the teacher finds out formally or informally what a student knows about a smaller, defined concept or skill and provides timely feedback to the student about his or her level of proficiency. Summative assessments occur when students have performed all activities in the module and are given a cumulative performance evaluation in which they demonstrate their growth in learning.

A comprehensive assessment system can be thought of as akin to a sporting event. Formative assessments are the practices: It is important to accomplish them consistently, they provide feedback to help students improve their learning, and making mistakes can be worthwhile if students are given an opportunity to learn from them. Summative assessments are the competitions: Students need to be prepared to perform at the best of their ability. Without multiple opportunities to practice skills along the way through formative assessments, students will not have the best chance of demonstrating growth in abilities through summative assessments (Black and Wiliam 1998).

## Embedded Formative Assessments

Formative assessments in this module serve two main purposes: to provide feedback to students about their learning and to provide important information for the teacher to inform immediate instructional needs. Providing feedback to students is particularly important when conducting problem- or project-based learning because students take on much of the responsibility for learning, and teachers must facilitate student learning in an informed way. For example, if students are required to conduct research for the Activity/Exploration phase but are not familiar with what constitutes a reliable resource, they may develop misconceptions based on poor information. When a teacher monitors this learning through formative assessments and provides specific feedback related to the instructional goals, students are less likely to develop incomplete or incorrect conceptions in their independent investigations. By using formative assessment to detect problems in student learning and then acting on this information, teachers help move student learning forward through these teachable moments.

Formative assessments come in a variety of formats. They can be informal, such as asking students probing questions related to student knowledge or tasks or simply

observing students engaged in an activity to gather information about student skills. Formative assessments can also be formal, such as a written quiz or a laboratory practical. Regardless of the type, three key steps must be completed when using formative assessments (Sondergeld, Bell, and Leusner 2010). First, the assessment is delivered to students so that teachers can collect data. Next, teachers analyze the data (student responses) to determine student strengths and areas that need additional support. Finally, teachers use the results from information collected to modify lessons and create learning environments that reinforce weak points in student learning. If student learning information is not used to modify instruction, the assessment cannot be considered formative in nature.

Formative assessments can be about content, science process skills, or even learning skills. When a formative assessment focuses on content, it assesses student knowledge about the disciplinary core ideas from the *Next Generation Science Standards* (*NGSS*) or content objectives from *Common Core State Standards for Mathematics* (*CCSS Mathematics*) or *Common Core State Standards for English Language Arts* (*CCSS ELA*). Content-focused formative assessments ask students questions about declarative knowledge regarding the concepts they have been learning. Process skills formative assessments examine the extent to which a student can perform science and engineering practices from the *NGSS* or process objectives from *CCSS Mathematics* or *CCSS ELA*, such as constructing an argument. Learning skills can also be assessed formatively by asking students to reflect on the ways they learn best during a module and identify ways they could have learned more.

## Assessment Maps

Assessment maps or blueprints can be used to ensure alignment between classroom instruction and assessment. If what students are learning in the classroom is not the same as the content on which they are assessed, the resultant judgment made on student learning will be invalid (Brookhart and Nitko 2008). Therefore, the issue of instruction and assessment alignment is critical. The assessment map for this book (found in Chapter 3) indicates by lesson whether the assessment should be completed as a group or on an individual basis, identifies the assessment as formative or summative in nature, and aligns the assessment with its corresponding learning objectives.

Note that the module includes far more formative assessments than summative assessments. This is done intentionally to provide students with multiple opportunities to practice their learning of new skills before completing a summative assessment. Note also that formative assessments are used to collect information on only one or two learning objectives at a time so that potential relearning or instructional modifications can focus on smaller and more manageable chunks of information. Conversely, summative assessments in the module cover many more learning objectives, as they are traditionally used as final markers of student learning. This is not to say that information collected from summative assessments cannot or should not be used formatively. If teachers find that gaps in student

learning persist after a summative assessment is completed, it is important to revisit these existing misconceptions or areas of weakness before moving on (Black et al. 2003).

## SELF-REGULATED LEARNING THEORY IN THE STEM ROAD MAP MODULES

Many learning theories are compatible with the STEM Road Map modules, such as constructivism, situated cognition, and meaningful learning. However, we feel that the self-regulated learning theory (SRL) aligns most appropriately (Zimmerman 2000). SRL requires students to understand that thinking needs to be motivated and managed (Ritchhart, Church, and Morrison 2011). The STEM Road Map modules are student centered and are designed to provide students with choices, concrete hands-on experiences, and opportunities to see and make connections, especially across subjects (Eliason and Jenkins 2012; NAEYC 2016). Additionally, SRL is compatible with the modules because it fosters a learning environment that supports students' motivation, enables students to become aware of their own learning strategies, and requires reflection on learning while experiencing the module (Peters and Kitsantas 2010).

The theory behind SRL (see Figure 2.2) explains the different processes that students engage in before, during, and after a learning task. Because SRL is a cyclical learning process, the accomplishment of one cycle develops strategies for the next learning cycle. This cyclic way of learning aligns with the various sections in the STEM Road Map lesson plans on Introductory Activity/ Engagement, Activity/Exploration, Explanation, Elaboration/Application of Knowledge, and Evaluation/Assessment. Since the students engaged in a module take on much of the responsibility for learning, this theory also provides guidance for teachers to keep students on the right track.

The remainder of this section explains how SRL theory is embedded within the five sections of each module and points out ways to

**Figure 2.2. SRL Theory**

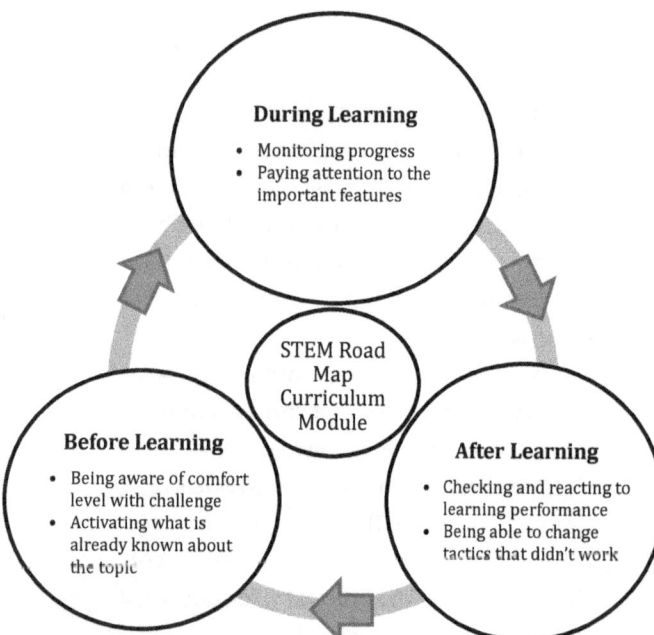

*Source:* Adapted from Zimmerman 2000.

support students in becoming independent learners of STEM while productively functioning in collaborative teams.

## Before Learning: Setting the Stage

Before attempting a learning task such as the STEM Road Map modules, teachers should develop an understanding of their students' level of comfort with the process of accomplishing the learning and determine what they already know about the topic. When students are comfortable with attempting a learning task, they tend to take more risks in learning and as a result achieve deeper learning (Bandura 1986).

The STEM Road Map curriculum modules are designed to foster excitement from the very beginning. Each module has an Introductory Activity/Engagement section that introduces the overall topic from a unique and exciting perspective, engaging the students to learn more so that they can accomplish the challenge. The Introductory Activity also has a design component that helps teachers assess what students already know about the topic of the module. In addition to the deliberate designs in the lesson plans to support SRL, teachers can support a high level of student comfort with the learning challenge by finding out if students have ever accomplished the same kind of task and, if so, asking them to share what worked well for them.

## During Learning: Staying the Course

Some students fear inquiry learning because they aren't sure what to do to be successful (Peters 2010). However, the STEM Road Map curriculum modules are embedded with tools to help students pay attention to knowledge and skills that are important for the learning task and to check student understanding along the way. One of the most important processes for learning is the ability for learners to monitor their own progress while performing a learning task (Peters 2012). The modules allow students to monitor their progress with tools such as the STEM Research Notebooks, in which they record what they know and can check whether they have acquired a complete set of knowledge and skills. The STEM Road Map modules support inquiry strategies that include previewing, questioning, predicting, clarifying, observing, discussing, and journaling (Morrison and Milner 2014). Through the use of technology throughout the modules, inquiry is supported by providing students access to resources and data while enabling them to process information, report the findings, collaborate, and develop 21st century skills.

It is important for teachers to encourage students to have an open mind about alternative solutions and procedures (Milner and Sondergeld 2015) when working through the STEM Road Map curriculum modules. Novice learners can have difficulty knowing what to pay attention to and tend to treat each possible avenue for information as equal (Benner 1984). Teachers are the mentors in a classroom and can point out ways for students to approach learning during the Activity/Exploration, Explanation, and

Elaboration/Application of Knowledge portions of the lesson plans to ensure that students pay attention to the important concepts and skills throughout the module. For example, if a student is to demonstrate conceptual awareness of motion when working on roller coaster research, but the student has misconceptions about motion, the teacher can step in and redirect student learning.

### After Learning: Knowing What Works

The classroom is a busy place, and it may often seem that there is no time for self-reflection on learning. Although skipping this reflective process may save time in the short term, it reduces the ability to take into account things that worked well and things that didn't so that teaching the module may be improved next time. In the long run, SRL skills are critical for students to become independent learners who can adapt to new situations. By investing the time it takes to teach students SRL skills, teachers can save time later, because students will be able to apply methods and approaches for learning that they have found effective to new situations. In the Evaluation/Assessment portion of the STEM Road Map curriculum modules, as well as in the formative assessments throughout the modules, two processes in the after-learning phase are supported: evaluating one's own performance and accounting for ways to adapt tactics that didn't work well. Students have many opportunities to self-assess in formative assessments, both in groups and individually, using the rubrics provided in the modules.

The designs of the *NGSS* and *CCSS* allow for students to learn in diverse ways, and the STEM Road Map curriculum modules emphasize that students can use a variety of tactics to complete the learning process. For example, students can use STEM Research Notebooks to record what they have learned during the various research activities. Notebook entries might include putting objectives in students' own words, compiling their prior learning on the topic, documenting new learning, providing proof of what they learned, and reflecting on what they felt successful doing and what they felt they still needed to work on. Perhaps students didn't realize that they were supposed to connect what they already knew with what they learned. They could record this and would be prepared in the next learning task to begin connecting prior learning with new learning.

## SAFETY IN STEM

Student safety is a primary consideration in all subjects but is an area of particular concern in science, where students may interact with unfamiliar tools and materials that may pose additional safety risks. It is important to implement safety practices within the context of STEM investigations, whether in a classroom laboratory or in the field. When you keep safety in mind as a teacher, you avoid many potential issues with the lesson while also protecting your students.

STEM safety practices encompass things considered in the typical science classroom. Ensure that students are familiar with basic safety considerations, such as wearing protective equipment (e.g., safety glasses or goggles and latex-free gloves) and taking care with sharp objects, and know emergency exit procedures. Teachers should learn beforehand the locations of the safety eyewash, fume hood, fire extinguishers, and emergency shut-off switch in the classroom and how to use them. Also be aware of any school or district safety policies that are in place and apply those that align with the work being conducted in the lesson. It is important to review all safety procedures annually.

STEM investigations should always be supervised. Each lesson in the modules includes teacher guidelines for applicable safety procedures that should be followed. Before each investigation, teachers should go over these safety procedures with the student teams. Some STEM focus areas such as engineering require that students can demonstrate how to properly use equipment in the maker space before the teacher allows them to proceed with the lesson.

The National Science Teaching Association (NSTA) provides a list of science rules and regulations, including standard operating procedures for lab safety, and a safety acknowledgment form for students and parents or guardians to sign. You can access these resources at *http://static.nsta.org/pdfs/SafetyInTheScienceClassroom.pdf*. In addition, NSTA's Safety in the Science Classroom web page (*www.nsta.org/safety*) has numerous links to safety resources, including papers written by the NSTA Safety Advisory Board.

*Disclaimer:* The safety precautions for each activity are based on use of the recommended materials and instructions, legal safety standards, and better professional practices. Using alternative materials or procedures for these activities may jeopardize the level of safety and therefore is at the user's own risk.

## REFERENCES

Bandura, A. 1986. *Social foundations of thought and action: A social cognitive theory.* Englewood Cliffs, NJ: Prentice-Hall.

Barell, J. 2006. *Problem-based learning: An inquiry approach.* Thousand Oaks, CA: Corwin Press.

Benner, P. 1984. *From novice to expert: Excellence and power in clinical nursing practice.* Menlo Park, CA: Addison-Wesley.

Black, P., C. Harrison, C. Lee, B. Marshall, and D. Wiliam. 2003. *Assessment for learning: Putting it into practice.* Berkshire, UK: Open University Press.

Black, P., and D. Wiliam. 1998. Inside the black box: Raising standards through classroom assessment. *Phi Delta Kappan* 80 (2): 139–148.

Blumenfeld, P., E. Soloway, R. Marx, J. Krajcik, M. Guzdial, and A. Palincsar. 1991. Motivating project-based learning: Sustaining the doing, supporting learning. *Educational Psychologist* 26 (3): 369–398.

Brookhart, S. M., and A. J. Nitko. 2008. *Assessment and grading in classrooms.* Upper Saddle River, NJ: Pearson.

Bybee, R., J. Taylor, A. Gardner, P. Van Scotter, J. Carlson Powell, A. Westbrook, and N. Landes. 2006. *The BSCS 5E instructional model: Origins and effectiveness.* Colorado Springs, CO: BSCS.

Eliason, C. F., and L. T. Jenkins. 2012. *A practical guide to early childhood curriculum.* 9th ed. New York: Merrill.

Johnson, C. 2003. Bioterrorism is real-world science: Inquiry-based simulation mirrors real life. *Science Scope* 27 (3): 19–23.

Krajcik, J., and P. Blumenfeld. 2006. Project-based learning. In *The Cambridge handbook of the learning sciences,* ed. R. Keith Sawyer, 317–334. New York: Cambridge University Press.

Lambros, A. 2004. *Problem-based learning in middle and high school classrooms: A teacher's guide to implementation.* Thousand Oaks, CA: Corwin Press.

Milner, A. R., and T. Sondergeld. 2015. Gifted urban middle school students: The inquiry continuum and the nature of science. *National Journal of Urban Education and Practice* 8 (3): 442–461.

Morrison, V., and A. R. Milner. 2014. Literacy in support of science: A closer look at cross-curricular instructional practice. *Michigan Reading Journal* 46 (2): 42–56.

National Association for the Education of Young Children (NAEYC). 2016. Developmentally appropriate practice position statements. *www.naeyc.org/positionstatements/dap.*

Peters, E. E. 2010. Shifting to a student-centered science classroom: An exploration of teacher and student changes in perceptions and practices. *Journal of Science Teacher Education* 21 (3): 329–349.

Peters, E. E. 2012. Developing content knowledge in students through explicit teaching of the nature of science: Influences of goal setting and self-monitoring. *Science and Education* 21 (6): 881–898.

Peters, E. E., and A. Kitsantas. 2010. The effect of nature of science metacognitive prompts on science students' content and nature of science knowledge, metacognition, and self-regulatory efficacy. *School Science and Mathematics* 110: 382–396.

Popham, W. J. 2013. *Classroom assessment: What teachers need to know.* 7th ed. Upper Saddle River, NJ: Pearson.

Ritchhart, R., M. Church, and K. Morrison. 2011. *Making thinking visible: How to promote engagement, understanding, and independence for all learners.* San Francisco, CA: Jossey-Bass.

Sondergeld, T. A., C. A. Bell, and D. M. Leusner. 2010. Understanding how teachers engage in formative assessment. *Teaching and Learning* 24 (2): 72–86.

Zimmerman, B. J. 2000. Attaining self-regulation: A social-cognitive perspective. In *Handbook of self-regulation,* ed. M. Boekaerts, P. Pintrich, and M. Zeidner, 13–39. San Diego: Academic Press.

# PART 2

# HUMAN IMPACTS ON OUR CLIMATE

## STEM ROAD MAP MODULE

# HUMAN IMPACTS ON OUR CLIMATE MODULE OVERVIEW

*Toni Ivey, Adrienne Redmond-Sanogo, Juliana Utley, Sue Christian Parsons, Janet B. Walton, Carla C. Johnson, and Erin Peters-Burton*

**THEME**: Cause and Effect

**LEAD DISCIPLINE**: Science

## MODULE SUMMARY

In sixth grade, students begin to grapple with some of the biggest challenges, and often debates, within and outside of the scientific community. In this module, science teachers take the lead, integrating with mathematics, social studies, and English language arts contexts, which could be collaborations with these classes. Students will investigate aspects of climate change driven by the rise in global temperatures over the past century and will develop potential solutions that might address one aspect of human activity that has contributed to global climate change. This project requires students to use an engineering design process to identify a problem and develop a method to help mitigate the identified problem (adapted from Johnson et al. 2015, p. 99).

## ESTABLISHED GOALS AND OBJECTIVES

At the conclusion of this module, students will be able to do the following:

- Explain the causes and effects of climate change and how humans have influenced climate change

- Understand how mathematical modeling and numerical data are used to determine the impacts of climate change

- Analyze and synthesize reputable media to form scientific arguments regarding climate change

- Describe the effects of climate change on the economy, society, and human populations

## CHALLENGE OR PROBLEM FOR STUDENTS TO SOLVE: THINK GLOBALLY, ACT LOCALLY CHALLENGE

Student teams are challenged to identify a local environmental problem and develop a method for monitoring and minimizing its impact on the environment. To support this goal, they learn about the differences between weather and climate and explore changes in temperature as an indicator of global warming. Students also investigate the role that greenhouse gases play in global warming.

**Driving Question:** How can we develop a local response to address an aspect of human impact on global climate change?

## CONTENT STANDARDS ADDRESSED IN THIS STEM ROAD MAP MODULE

A full listing with descriptions of the standards this module addresses can be found in the appendix. Listings of the particular standards addressed within lessons are provided in a table for each lesson in Chapter 4.

## STEM RESEARCH NOTEBOOK

Each student should maintain a STEM Research Notebook, which will serve as a place for students to organize their work throughout this module (see p. 12 for more general discussion on setup and use of this notebook). All written work in the module should be included in the notebook, including records of students' thoughts and ideas, fictional accounts based on the concepts in the module, and records of student progress through the module's engineering design process (EDP). The notebooks may be maintained across subject areas, giving students the opportunity to see that although their classes may be separated during the school day, the knowledge they gain is connected.

Each lesson in this module includes student handouts that should be kept in the STEM Research Notebooks after completion, as well as prompts to which students should respond in their notebooks. You may also wish to have students include the STEM Research Notebook Guidelines student handout in their notebooks.

Emphasize to students the importance of organizing all information in a Research Notebook. Explain that scientists and other researchers maintain detailed Research Notebooks in their work. These notebooks, which are crucial to researchers' work because they contain critical information and track the researchers' progress, are often considered legal documents for scientists who are pursuing patents or wish to provide proof of their discovery process.

**STUDENT HANDOUT**

# STEM RESEARCH NOTEBOOK GUIDELINES

STEM professionals record their ideas, inventions, experiments, questions, observations, and other work details in notebooks so that they can use these notebooks to help them think about their projects and the problems they are trying to solve. You will each keep a STEM Research Notebook during this module that is like the notebooks that STEM professionals use. In this notebook, you will include all your work and notes about ideas you have. The notebook will help you connect your daily work with the big problem or challenge you are working to solve.

It is important that you organize your notebook entries under the following headings:

1. **Chapter Topic or Title of Problem or Challenge:** You will start a new chapter in your STEM Research Notebook for each new module. This heading is the topic or title of the big problem or challenge that your team is working to solve in this module.

2. **Date and Topic of Lesson Activity for the Day:** Each day, you will begin your daily entry by writing the date and the day's lesson topic at the top of a new page. Write the page number both on the page and in the table of contents.

3. **Information Gathered From Research:** This is information you find from outside resources such as websites or books.

4. **Information Gained From Class or Discussions With Team Members:** This information includes any notes you take in class and notes about things your team discusses. You can include drawings of your ideas here, too.

5. **New Data Collected From Investigations:** This includes data gathered from experiments, investigations, and activities in class.

6. **Documents:** These are handouts and other resources you may receive in class that will help you solve your big problem or challenge. Paste or staple these documents in your STEM Research Notebook for safekeeping and easy access later.

7. **Personal Reflections:** Here, you record your own thoughts and ideas on what you are learning.

8. **Lesson Prompts:** These are questions or statements that your teacher assigns you within each lesson to help you solve your big problem or challenge. You will respond to the prompts in your notebook.

9. **Other Items:** This section includes any other items your teacher gives you or other ideas or questions you may have.

## MODULE LAUNCH

To launch the module, introduce the Think Globally, Act Locally Challenge by informing students that their challenge in this module will be to develop a method for monitoring and minimizing a human activity that has an impact on the environment and has contributed to global climate change. To do so, they will first learn about weather and climate, global warming, climate change, and climate change indicators. Then, they will identify a local environmental problem and develop their own solutions to address the challenge. This challenge will require them to conduct research on the causes of climate change, interview experts and others with understanding of this topic, and use mathematical modeling and numerical data to determine what steps have been taken to mitigate climate change.

## PREREQUISITE SKILLS FOR THE MODULE

Students enter this module with a wide range of preexisting skills, information, and knowledge. Table 3.1 provides an overview of prerequisite skills and knowledge that students are expected to apply in this module, along with examples of how they apply this knowledge throughout the module. Differentiation strategies are also provided for students who may need additional support in acquiring or applying this knowledge.

**Table 3.1. Prerequisite Key Knowledge and Examples of Applications and Differentiation Strategies**

| Prerequisite Key Knowledge | Application of Knowledge | Differentiation for Students Needing Additional Support |
|---|---|---|
| *Science*<br>• Organize data into appropriate tables, graphs, drawings, or diagrams.<br>• Have knowledge of basic weather and different types of climates. | *Science*<br>• Create tables, graphs, and charts of climate data.<br>• Differentiate between weather and climate data. | *Science*<br>• Preselect data to assist students in collecting data.<br>• Provide opportunities for students to practice creating graphs.<br>• Provide templates for students to fill out with collected information so that the data can be transferred to their STEM Research Notebooks.<br>• Provide students with handouts to help them develop their multimedia presentations.<br>• Provide access to texts and media for students to learn about weather and climate.<br>• Include a lesson on weather if needed. |

*Continued*

**Table 3.1.** (*continued*)

| Prerequisite Key Knowledge | Application of Knowledge | Differentiation for Students Needing Additional Support |
|---|---|---|
| *Mathematics*<br>Numbers and Operations:<br>• Multiply a whole number of up to four digits by a one-digit whole number, and multiply two two-digit numbers, using strategies based on place value and the properties of operations.<br>• Illustrate and explain the calculation by using equations, rectangular arrays, and/or area models.<br>• Solve word problems involving multiplication of a fraction by a whole number.<br><br>The Number System:<br>• Solve real-world and mathematical problems by graphing points in all four quadrants of the coordinate plane.<br>• Include use of coordinates and absolute value to find distances between points with the same first coordinate or the same second coordinates.<br><br>Measurement and Data:<br>• Draw a scaled picture graph and a scaled bar graph to represent a data set with several categories.<br>• Know relative sizes of measurement units within one system of units including km, m, cm; kg, g; lb, oz.; l, ml; hr, min, sec.<br>• Apply the area and perimeter formulas for rectangles in real world and mathematical problems. | *Mathematics*<br>Numbers and Operations:<br>• Use whole number operations and multiplication and division of fractions to solve problems in the module.<br>• Understand the area of a rectangle with whole number and fractional side measurements in order to understand the size of an acre.<br><br>The Number System:<br>• Understand how to read and plot points in the four quadrants of a Cartesian plane in order to create and interpret graphs.<br><br>Measurement and Data:<br>• Measure temperature and volumes.<br>• Read graphs and create graphs.<br>• Collect data from databases.<br>• Understand the concepts of mean, median, and mode. | *Mathematics*<br>Numbers and Operations:<br>• Provide calculators.<br>• Allow students to have access to hundreds grids, physical manipulatives, and other representations so that they can work with fractions and decimals.<br><br>The Number System:<br>• Use anchor charts and other representations to help students understand coordinate grids.<br><br>Measurement and Data:<br>• Work with students individually or pair them with other students to help them understand the data and measurement objectives.<br>• Provide digital thermometers if necessary. |

*Continued*

**Table 3.1.** (*continued*)

| Prerequisite Key Knowledge | Application of Knowledge | Differentiation for Students Needing Additional Support |
|---|---|---|
| *Reading*<br>• Have critical reading skills, including making inferences, citing textual evidence, and summarizing central ideas. | *Reading*<br>• Engage with a wide variety of texts.<br>• Analyze and evaluate data and argument.<br>• Analyze nonfiction texts for purpose, structures, and features, and use learned comprehension strategies. | *Reading*<br>• Take care to include texts reflecting a range of reading levels.<br>• Use nonfiction trade books with photographs, illustrations, and graphics to support student understanding of text. |
| *Writing*<br>• Write arguments to support claims in an analysis of substantive topics or texts, using valid reasoning and relevant and sufficient evidence.<br>• Use technology, including the internet, to interact and collaborate with others. | *Writing*<br>• Explore background knowledge and grow understandings, take notes in the process of research, and integrate and share information recorded in this way.<br>• Create and share presentations and develop action plans.<br>• Apply developing strategies related to nonfiction writing.<br>• Develop and strengthen writing as needed by planning, revising, editing, rewriting, or trying a new approach. | *Writing*<br>• Provide writing skill support as necessary. |
| *Research and Communication*<br>• Clearly present information, findings, and supporting evidence in a manner appropriate to the task, purpose, and audience.<br>• Make strategic use of digital media and visual displays of data to express information and enhance understanding of presentation. | *Research and Communication*<br>• Prepare and give presentations to peers or to invited members of the community and parents. | *Research and Communication*<br>• Monitor and support learners as they work to express their ideas.<br>• Actively support student presentation skills as needed. |

## POTENTIAL STEM MISCONCEPTIONS

Students enter the classroom with a wide variety of prior knowledge and ideas, so it is important to be alert to misconceptions, or inappropriate understandings of foundational knowledge. These misconceptions can be classified as one of several types: "preconceived notions," opinions based on popular beliefs or understandings; "nonscientific beliefs," knowledge students have gained about science from sources outside the scientific community; "conceptual misunderstandings," incorrect conceptual models based on incomplete understanding of concepts; "vernacular misconceptions," misunderstandings of words based on their common use versus their scientific use; and "factual misconceptions," incorrect or imprecise knowledge learned in early life that remains unchallenged (NRC 1997, p. 28). Misconceptions must be addressed and dismantled in order for students to reconstruct their knowledge, and therefore teachers should be prepared to take the following steps:

- *Identify students' misconceptions.*

- *Provide a forum for students to confront their misconceptions.*

- *Help students reconstruct and internalize their knowledge, based on scientific models.*
  *(NRC 1997, p. 29)*

Keeley and Harrington (2010) recommend using diagnostic tools such as probes and formative assessment to identify and confront student misconceptions and begin the process of reconstructing student knowledge. Keeley's *Uncovering Student Ideas in Science* series contains probes targeted toward uncovering student misconceptions in a variety of areas and may be a useful resource for addressing student misconceptions in this module.

Some commonly held misconceptions specific to lesson content are provided with each lesson so that you can be alert for student misunderstanding of the science concepts presented and used during this module. The American Association for the Advancement of Science has also identified misconceptions that students frequently hold regarding various science concepts (see the links at *http://assessment.aaas.org/topics*).

## SRL PROCESS COMPONENTS

Table 3.2 (p. 30) illustrates some of the activities in the Human Impacts on Our Climate module and how they align with the self-regulated learning (SRL) process before, during, and after learning.

**Table 3.2.** SRL Process Components

| Learning Process Components | Example From Human Impacts on Our Climate Module | Lesson Number and Learning Component |
|---|---|---|
| **BEFORE LEARNING** | | |
| Motivates students | Students participate in a gallery walk of pictures they have created of their perception of the environment. | Lesson 1, Introductory Activity/ Engagement |
| Evokes prior learning | Students tap into their prior experience with weather and apply it to distinguish between weather and climate. | Lesson 1, Activity/ Exploration |
| Helps students monitor their progress | Students reflect on the use of biased or misleading data in their personal decision-making processes and consider their own understanding of climate change in this context. | Lesson 1, Elaboration/ Application of Knowledge |
| **DURING LEARNING** | | |
| Focuses on important features | Students create models of greenhouses to explore the effects of greenhouse gases on Earth. | Lesson 2, Activity/ Exploration |
| Helps students monitor their progress | Students respond to STEM Research Notebook prompts to check their understanding of climate change indicators and evidence for climate change. | Lesson 2, Elaboration/ Application of Knowledge |
| **AFTER LEARNING** | | |
| Evaluates learning | In the final challenge, students present their solutions to a problem and obtain feedback. | Lesson 3, Elaboration/ Application of Knowledge |
| Takes account of what worked and what did not work | Students complete post-tests and reflect on their mitigation plans. | Lesson 3, Elaboration/ Application of Knowledge |

## STRATEGIES FOR DIFFERENTIATING INSTRUCTION WITHIN THIS MODULE

For the purposes of this curriculum module, differentiated instruction is conceptualized as a way to tailor instruction—including process, content, and product—to various student needs in your class. A number of differentiation strategies are integrated into

lessons across the module. The problem- and project-based learning approach used in the lessons is designed to address students' multiple intelligences by providing a variety of entry points and methods to investigate the key concepts in the module. Differentiation strategies for students needing support in prerequisite knowledge can be found in Table 3.1 (p. 26). You are encouraged to use information gained about student prior knowledge during introductory activities and discussions to inform your instructional differentiation. Strategies incorporated into this lesson include flexible grouping, varied environmental learning contexts, assessments, compacting, and tiered assignments and scaffolding.

*Flexible Grouping.* Students work collaboratively in a variety of activities throughout this module. Grouping strategies you might employ include student-led grouping, grouping students according to ability level or common interests, grouping students randomly, or grouping them so that students in each group have complementary strengths (for instance, one student might be strong in mathematics, another in art, and another in writing).

*Varied Environmental Learning Contexts.* Students have the opportunity to learn in various contexts throughout the module, including alone, in groups, in quiet reading and research-oriented activities, and in active learning through inquiry and design activities. In addition, students learn in a variety of ways, including through doing inquiry activities, journaling, reading fiction and nonfiction texts, watching videos, participating in class discussion, and conducting web-based research.

*Assessments.* Students are assessed in a variety of ways throughout the module, including individual and collaborative formative and summative assessments. Students have the opportunity to produce work via written text, oral and media presentations, and modeling. You may choose to provide students with additional choices of media for their products (for example, PowerPoint presentations, posters, or student-created websites or blogs).

*Compacting.* Based on student prior knowledge, you may wish to adjust instructional activities for students who exhibit prior mastery of a learning objective. For instance, if some students exhibit a pre-existing understanding of the differences between weather and climate in Lesson 1, you may wish to limit the amount of time they spend on learning this content and instead introduce ELA or social studies connections with associated activities.

*Tiered Assignments and Scaffolding.* Based on your awareness of student ability, understanding of concepts, and mastery of skills, you may wish to provide students with variations on activities by adding complexity to assignments or providing more or fewer learning supports for activities throughout the module. For instance, some students may need additional support in identifying key search words and phrases for web-based research or may benefit from cloze sentence handouts to enhance vocabulary

understanding. Other students may benefit from expanded reading selections and additional reflective writing or from working with manipulatives and other visual representations of mathematical concepts. You may also work with your school librarian to compile a set of topical resources at a variety of reading levels.

## STRATEGIES FOR ENGLISH LANGUAGE LEARNERS

Students who are developing proficiency in English language skills require additional supports to simultaneously learn academic content and the specialized language associated with specific content areas. WIDA (2012) has created a framework for providing support to these students and makes available rubrics and guidance on differentiating instructional materials for English language learners (ELLs). In particular, ELL students may benefit from additional sensory supports such as images, physical modeling, and graphic representations of module content, as well as interactive support through collaborative work.

When differentiating instruction for ELL students, you should carefully consider the needs of these students as you introduce and use academic language in various language domains (listening, speaking, reading, and writing) throughout this module. To adequately differentiate instruction for ELL students, you should have an understanding of the proficiency level of each student. The following five overarching WIDA learning standards are relevant to this module:

- Standard 1: Social and Instructional Language. Focus on social behavior in group work and class discussions.

- Standard 2: The Language of Language Arts. Focus on forms of print, elements of text, picture books, comprehension strategies, main ideas and details, persuasive language, creation of informational text, and editing and revision.

- Standard 3: The Language of Mathematics. Focus on numbers and operations, patterns, number sense, measurement, and strategies for problem solving.

- Standard 4: The Language of Science. Focus on safety practices, scientific process, and scientific inquiry.

- Standard 5: The Language of Social Studies. Focus on resources and environmental issues.

## SAFETY CONSIDERATIONS FOR THE ACTIVITIES IN THIS MODULE

For precautions, see the specific safety notes after the list of materials in the first two lessons. For more general safety guidelines, see the Safety in STEM section in Chapter 2

(p. 18). We also recommend that you go over the safety rules that are included as part of the safety acknowledgment form with your students before beginning the first investigation. Once you have gone over these rules with your students, have them sign the safety acknowledgment form. You should also send the form home with students for parents or guardians to read and sign to acknowledge that they understand the safety procedures that must be followed by their children. A sample middle school safety acknowledgment form can be found on the NSTA Safety Portal at *http://static.nsta.org/pdfs/SafetyAcknowledgmentForm-MiddleSchool.pdf.*

## DESIRED OUTCOMES AND MONITORING SUCCESS

The desired outcomes for this module are outlined in Table 3.3, along with suggested ways to gather evidence to monitor student success. For more specific details on desired outcomes, see the Established Goals and Objectives sections for the module (p. 23) and individual lessons.

**Table 3.3.** Desired Outcome and Evidence of Success in Achieving Identified Outcome

| Desired Outcome | Evidence of Success | |
| --- | --- | --- |
| | Performance Tasks | Other Measures |
| Students create and present a solution to a problem illustrating their understanding of the causes and effects of human impacts on the environment. | • Students are assessed on their presentations and their written descriptions of their solutions to reduce human impacts on the environment.<br>• Students maintain STEM Research Notebooks that contain designs, research notes, evidence of collaboration, and mathematics, social studies, and ELA-related work. | Students are assessed on the following:<br>• Collaboration in their groups.<br>• Use of claim, evidence, reasoning responses to assess changes in their understandings.<br>• Climate change pre- and post-tests.<br>• Participation in classroom discussions. |

## ASSESSMENT PLAN OVERVIEW AND MAP

Table 3.4 (p. 34) provides an overview of the major group and individual products and deliverables, or things that student teams will produce in this module, that constitute the assessment for this module. See Table 3.5 (p. 34) for a full assessment map of formative and summative assessments in this module.

**Table 3.4.** Major Products and Deliverables for Groups and Individuals

| Lesson | Major Group Products and Deliverables | Major Individual Products and Deliverables |
|---|---|---|
| 1 | • Graphs of changes in average global temperatures | • STEM Research Notebook entries |
| 2 | • Greenhouse Model Activity Sheet | • Greenhouse Effect Simulation Student Handout<br>• STEM Research Notebook entries |
| 3 | • Plan to save the bees | • STEM Research Notebook entries<br>• Think Globally, Act Locally presentation and paper |

**Table 3.5.** Assessment Map for Human Impacts on Our Climate Module

| Lesson | Assessment | Group/ Individual | Formative/ Summative | Lesson Objective Assessed |
|---|---|---|---|---|
| 1 | Global temperature trends *activity* | Group | Formative | • Analyze data and interpret trends in changes in average global temperature. |
| 1 | STEM Research Notebook *prompts* | Individual | Formative | • Identify and describe the layers of Earth's atmosphere.<br>• Use scientific data to form an informed position on changes in average global temperatures.<br>• Describe the differences between weather and climate. |
| 2 | Greenhouse model *activity* | Group | Formative | • Collect and analyze data to study the greenhouse effect. |

*Continued*

**Table 3.5.** (*continued*)

| Lesson | Assessment | Group/ Individual | Formative/ Summative | Lesson Objective Assessed |
|---|---|---|---|---|
| 2 | Greenhouse effect *simulation* | Group or individual | Summative | • Determine whether all greenhouse gases contribute to the greenhouse effect.<br>• Collect and analyze data to determine which gases are greenhouse gases.<br>• Make conclusions about the role of greenhouse gases in Earth's atmosphere, consider how greenhouse gases affect our climate, and predict what will happen to average global temperature. |
| 2 | STEM Research Notebook *prompt* | Individual | Formative | • Use scientific data to form an informed position on changes in Earth's climate. |
| 3 | STEM Research Notebook *prompts* | Individual | Formative | • Calculate their own and their households' carbon footprints.<br>• Describe the importance of reducing their carbon footprints.<br>• Understand how climate change affects other organisms and why this is important to humans. |
| 3 | Think Globally, Act Locally *presentation and paper* | Group or individual | Summative | • Synthesize their understanding of climate change to identify a climate concern in their own lives or in their school or community.<br>• Design a solution to reduce the carbon footprint associated with the identified climate concern. |

## MODULE TIMELINE

Tables 3.6–3.10 (pp. 37–40) provide lesson timelines for each week of the module. These timelines are provided for general guidance only and are based on class times of approximately 45 minutes.

**Table 3.6. STEM Road Map Module Schedule for Week One**

| Day 1 | Day 2 | Day 3 | Day 4 | Day 5 |
|---|---|---|---|---|
| *Lesson 1* *Weather Versus Climate and Global Warming Trends* | *Lesson 1* *Weather Versus Climate and Global Warming Trends* | *Lesson 1* *Weather Versus Climate and Global Warming Trends* | *Lesson 1* *Weather Versus Climate and Global Warming Trends* | *Lesson 2* *The Greenhouse Effect and Climate Change* |
| • Launch the module by introducing the challenge. | • Class discusses weather. | • Class discusses learnings from the video about weather. | • Student groups graph changes in average global temperatures and reflect on trends. | • Class discusses how everyday items insulate body for warmth. |
| • Students complete pre-test and drawings of environment, then hold gallery walk. | • Students watch a video explaining weather and make notes on a T-chart. | • Students watch a video about the layers of Earth's atmosphere and respond to a prompt in their STEM Research Notebooks, then make a Layers of Earth's Atmosphere Foldable. | • Students use claim, evidence, reasoning model to summarize the trend of data over the last century. | • Student teams build and test greenhouse models. |
| • Students write questions about climate change on exit tickets. | • Students explore the local 5-day weather forecast and make predictions using the claim, evidence, reasoning model. | • Students research human inventions and how they transformed society. | • Students continue to learn about weather versus climate by watching a video, adding to their T-charts, and responding to a prompt in their notebooks. | • Students learn about tides and sea level rise. |
| • Students explore the average temperature by month for various U.S. cities. | | | • Students construct a scatter plot and extrapolate from the data. | • Students investigate the use of persuasive language in advertising related to climate change. |
| | | | • Students explore average monthly temperatures. | • Students learn about biased versus unbiased data through simulations. |
| | | | • Students research engineering and compare it to inventing, and research the creation of an object of their choice. | |

**Table 3.7. STEM Road Map Module Schedule for Week Two**

| Day 6 | Day 7 | Day 8 | Day 9 | Day 10 |
|---|---|---|---|---|
| *Lesson 2* | *Lesson 2* | *Lesson 2* | *Lesson 2* | *Lesson 2* |
| *The Greenhouse Effect and Climate Change* | *The Greenhouse Effect and Climate Change* | *The Greenhouse Effect and Climate Change* | *The Greenhouse Effect and Climate Change* | *The Greenhouse Effect and Climate Change* |
| • Students work on greenhouse simulation, Parts 1–2. | • Students work on greenhouse simulation, Parts 3–4. | • Students work on greenhouse simulation, Parts 4–5. | • Students watch videos on greenhouse effect and causes of climate change. | • Students complete climate change indicator jigsaw activity. |
| • Students complete exit ticket on which gases are greenhouse gases. | • Students propose possible solutions for helping communities deal with the effects of climate change and create multimedia presentations. | • Students consider how the data from their greenhouse models correlate with what they learned from the simulation. | • Students begin climate change indicator jigsaw activity. | • Students respond to prompt in STEM Research Notebook. |
| | | • Students connect to the politics of climate change through studying current events. | • Students share their multimedia presentations. | • Small groups develop a claim, evidence, reasoning response based on data from NOAA indicator graphs and discuss the influence on our climate. |
| | | • Students examine how the Industrial Revolution changed society and how industry is continually revolutionizing. | | • Students investigate how climate change affects different sectors of society. |

**Table 3.8. STEM Road Map Module Schedule for Week Three**

| Day 11 | Day 12 | Day 13 | Day 14 | Day 15 |
|---|---|---|---|---|
| *Lesson 3* Reducing Your Carbon Footprint | *Lesson 3* Reducing Your Carbon Footprint | *Lesson 3* Reducing Your Carbon Footprint | *Lesson 3* Reducing Your Carbon Footprint | *Lesson 3* Reducing Your Carbon Footprint |
| • Students watch a video about the impact of climate change on bees, and then respond to related questions in their STEM Research Notebooks. | • Read aloud several books on bees.<br><br>• Students learn about bees around the world.<br><br>• Students read and analyze the Entomology Society of America's (ESA) position statement on climate change and answer questions about the accompanying figures. | • Students complete analysis of the ESA position statement.<br><br>• Teams list all the foods their team members eat regularly and consider which rely on bees.<br><br>• Students evaluate the reliability of digital resources.<br><br>• Teams create plans for saving the bees from extinction.<br><br>• Students begin work on visual presentation on causes and effects of climate change. | • Students continue work on climate change visual presentation.<br><br>• Students perform more calculations with the percentages of their food consumption that relies on bees.<br><br>• Students discuss and analyze how persuasive writing is used in advertising related to climate change. | • Teams make visual presentations on climate change.<br><br>• Students discuss issues that the death of the bees would cause and how changes in climate might change human activity. |

**Table 3.9. STEM Road Map Module Schedule for Week Four**

| Day 16 | Day 17 | Day 18 | Day 19 | Day 20 |
|---|---|---|---|---|
| *Lesson 3 Reducing Your Carbon Footprint*<br>• Introduce final challenge.<br>• Students calculate individual carbon footprints and compare with those of classmates.<br>• Students document ways they can reduce their carbon footprints. | *Lesson 3 Reducing Your Carbon Footprint*<br>• Students identify and research a local environmental problem.<br>• Read aloud *Energy Island: How One Community Harnessed the Wind and Changed Their World,* by Allan Drummond. | *Lesson 3 Reducing Your Carbon Footprint*<br>• Students research and discuss actions aimed at reducing carbon emissions.<br>• Students brainstorm solutions and pick the best one.<br>• Students determine any constraints to the plan. | *Lesson 3 Reducing Your Carbon Footprint*<br>• Students identify resources for plan and begin to interview experts. | *Lesson 3 Reducing Your Carbon Footprint*<br>• Students continue to interview experts. |

**Table 3.10. STEM Road Map Module Schedule for Week Five**

| Day 21 | Day 22 | Day 23 | Day 24 | Day 25 |
|---|---|---|---|---|
| *Lesson 3 Reducing Your Carbon Footprint*<br>• Students begin to develop and test mitigation plan. | *Lesson 3 Reducing Your Carbon Footprint*<br>• Students continue to develop and test mitigation plan. | *Lesson 3 Reducing Your Carbon Footprint*<br>• Students continue to develop, test, and revise mitigation plan. | *Lesson 3 Reducing Your Carbon Footprint*<br>• Students present final projects. | *Lesson 3 Reducing Your Carbon Footprint*<br>• Students write final paper.<br>• Students take post-test and make new environment drawing and compare with the pre-test and earlier drawing.<br>• Students hold a gallery walk.<br>• Students discuss how their ideas have changed since beginning the module. |

## RESOURCES

The media specialist can help you locate resources for students to view and read about weather, climate, environmental issues, climate change, and related content. Special educators and reading specialists can help find supplemental sources for students needing extra support in reading and writing. Additional resources may be found online. Community resources for this module may include community members for students to interview (e.g., city board members, city utilities and parks and recreation representatives, and school administrators), as well as meteorologists, climate scientists, and representatives of the state department of wildlife and fisheries.

## REFERENCES

Johnson, C. C., T. J. Moore, J. Utley, J. Breiner, S. R. Burton, E. E. Peter-Burton, J. Walton, and C. L. Parton. 2015. The STEM Road Map for grades 6–8. In *STEM Road Map: A framework for integrated STEM education,* ed. C. C. Johnson, E. E. Peters-Burton, and T. J. Moore, 96–123. New York: Routledge. *www.routledge.com/products/9781138804234.*

Keeley, P., and R. Harrington. 2010. *Uncovering student ideas in physical science, volume 1: 45 new force and motion assessment probes.* Arlington, VA: NSTA Press.

National Research Council (NRC). 1997. *Science teaching reconsidered: A handbook.* Washington, DC: National Academies Press.

WIDA. 2012. 2012 amplification of the English language development standards: Kindergarten–grade 12. *https://wida.wisc.edu/teach/standards/eld.*

# HUMAN IMPACTS ON OUR CLIMATE LESSON PLANS

*Toni Ivey, Adrienne Redmond-Sanogo, Juliana Utley, Sue Christian Parsons, Janet B. Walton, Carla C. Johnson, and Erin Peters-Burton*

## Lesson Plan 1: Weather Versus Climate and Global Warming Trends

This lesson builds foundational knowledge that will be used in the Think Globally, Act Locally Challenge, with a focus on current scientific understandings regarding global warming and climate change. In this lesson, students learn about weather and climate, including how climate differs from weather. Students graph and analyze data on changes in average global temperatures collected since 1880. From this graphing study, students make a claim about global warming trends based on their provided evidence.

### ESSENTIAL QUESTIONS

- What is weather?

- What is climate?

- How does weather differ from climate?

- Where does weather take place in Earth's atmosphere?

### ESTABLISHED GOALS AND OBJECTIVES

At the conclusion of this lesson, students will be able to do the following:

- Describe the differences between weather and climate

- Identify different types of daily weather

- Identify and describe the different layers of Earth's atmosphere

- Apply research skills to locate historical and current global temperature data

- Analyze data and interpret trends in changes in average global temperature

- Use scientific data to form an informed position on changes in average global temperatures

## TIME REQUIRED

- 4 days (approximately 45 minutes each day; see Table 3.6, p. 37)

## MATERIALS

*Required Materials for Lesson 1*

- STEM Research Notebooks (1 per student; see p. 25 for STEM Research Notebook student handout)
- Computer and projector for students to watch videos
- Computers, tablets, or laptops with internet access for student research
- Handouts (attached at the end of this lesson)
- Butcher paper or large piece of paper for poster (1 per group)
- Markers (several per group)
- 3" × 5" index cards (1 per student)
- Graph paper
- Tape to hang drawings and timelines on wall
- Scissors (1 pair per student)
- Ruler (1 per student)
- Stapler to staple foldable together (1 per 4 students)

## SAFETY NOTE

- Use caution when working with scissors, which can be sharp and can cut or puncture skin.

## CONTENT STANDARDS AND KEY VOCABULARY

Table 4.1 lists the content standards from the *Next Generation Science Standards* (NGSS), *Common Core State Standards* (CCSS), and the Framework for 21st Century Learning that this lesson addresses, and Table 4.2 (p. 48) presents the key vocabulary. Vocabulary terms are provided for both teacher and student use. Teachers may choose to introduce some or all of the terms to students.

**Table 4.1.** Content Standards Addressed in STEM Road Map Module
Lesson 1

---

*NEXT GENERATION SCIENCE STANDARDS*

PERFORMANCE EXPECTATIONS
- MS-ESS3-5. Ask questions to clarify evidence of the factors that have caused the rise in global temperatures over the past century.

SCIENCE AND ENGINEERING PRACTICES

*Asking Questions and Defining Problems*

Asking questions and defining problems in grades 6–8 builds on grades K–5 experiences and progresses to specifying relationships between variables, and clarifying arguments and models.
- Ask questions to identify and clarify evidence of an argument.

*Analyzing and Interpreting Data*

Analyzing data in 6–8 builds on K–5 and progresses to extending quantitative analysis to investigations, distinguishing between correlation and causation, and basic statistical techniques of data and error analysis.
- Analyze and interpret data to determine similarities and differences in findings.

*Constructing Explanations and Designing Solutions*

Constructing explanations and designing solutions in 6–8 builds on K–5 experiences and progresses to include constructing explanations and designing solutions supported by multiple sources of evidence consistent with scientific ideas, principles, and theories.
- Apply scientific principles to design an object, tool, process or system.

*Engaging in Argument From Evidence*

Engaging in argument from evidence in 6–8 builds on K–5 experiences and progresses to constructing a convincing argument that supports or refutes claims for either explanations or solutions about the natural and designed world(s).
- Construct an oral and written argument supported by empirical evidence and scientific reasoning to support or refute an explanation or a model for a phenomenon or a solution to a problem.

DISCIPLINARY CORE IDEAS

*ESS2.D. Weather and Climate*
- Because these patterns are so complex, weather can only be predicted probabilistically.

*ESS3.C. Human Impacts on Earth Systems*
- Human activities have significantly altered the biosphere, sometimes damaging or destroying natural habitats and causing the extinction of other species. But changes to Earth's environments can have different impacts (negative and positive) for different living things.

---

*Continued*

**Table 4.1.** (*continued*)

> • Typically as human populations and per-capita consumption of natural resources increase, so do the negative impacts on Earth, unless the activities and technologies involved are engineered otherwise.
>
> ## CROSSCUTTING CONCEPTS
>
> *Patterns*
>
> • Graphs, charts, and images can be used to identify patterns in data.
>
> *Cause and Effect*
>
> • Relationships can be classified as causal or correlational, and correlation does not necessarily imply causation.
>
> • Cause and effect relationship may be used to predict phenomena in natural or designed systems.
>
> *Stability and Change*
>
> • Stability might be disturbed either by sudden events or gradual changes that accumulate over time.
>
> ## COMMON CORE STATE STANDARDS FOR MATHEMATICS
>
> ### MATHEMATICAL PRACTICES
>
> • MP1. Make sense of problems and persevere in solving them.
>
> • MP2. Reason abstractly and quantitatively.
>
> • MP3. Construct viable arguments and critique the reasoning of others.
>
> • MP4. Model with mathematics.
>
> • MP5. Use appropriate tools strategically.
>
> • MP6. Attend to precision.
>
> • MP8. Look for and express regularity in repeated reasoning.
>
> ### MATHEMATICAL CONTENT
>
> • 6.SP.A.3. Recognize that a measure of center for a numerical data set summarizes all of its values with a single number, while a measure of variation describes how its values vary with a single number.
>
> • 6.SP.5. Summarize numerical data sets in relation to their context.
>
> ## COMMON CORE STATE STANDARDS FOR ENGLISH LANGUAGE ARTS
>
> ### READING STANDARDS
>
> • RI.6.1. Cite textual evidence to support analysis of what the text says explicitly as well as inferences drawn from the text.

*Continued*

**Table 4.1.** (*continued*)

- RI.6.2. Determine a central idea of a text and how it is conveyed through particular details; provide a summary of the text distinct from personal opinions or judgments.

- RI.6.7. Integrate information presented in different media or formats (e.g., visually, quantitatively) as well as in words to develop a coherent understanding of a topic or issue.

## WRITING STANDARDS

- WHST.6-8.2.D. Use precise language and domain-specific vocabulary to inform about or explain the topic.

- WHST.6-8.8. Gather relevant information from multiple print and digital sources, using search terms effectively; assess the credibility and accuracy of each source; and quote or paraphrase the data and conclusions of others while avoiding plagiarism and following a standard format for citation.

## SPEAKING AND LISTENING STANDARDS

- SL.6.1. Engage effectively in a range of collaborative discussions (one-on-one, in groups, and teacher-led) with diverse partners on grade 6 topics, texts, and issues, building on others' ideas and expressing their own clearly.

- SL.6.1.A. Come to discussions prepared, having read or studied required material; explicitly draw on that preparation by referring to evidence on the topic, text, or issue to probe and reflect on ideas under discussion.

- SL.6.1.B. Follow rules for collegial discussions, set specific goals and deadlines, and define individual roles as needed.

- SL.6.1.C. Pose and respond to specific questions with elaboration and detail by making comments that contribute to the topic, text, or issue under discussion.

## FRAMEWORK FOR 21ST CENTURY LEARNING

- Interdisciplinary Themes: Global Awareness, Environmental Literacy

- Learning and Innovation Skills: Critical Thinking and Problem Solving, Communication, Collaboration

- Information, Media, and Technology Skills: Information Literacy

- Life and Career Skills: Productivity and Accountability

**Table 4.2. Key Vocabulary for Lesson 1**

| Key Vocabulary | Definition |
|---|---|
| abiotic | nonliving; not biological |
| climate | the average weather conditions over a long period (typically no less than 30 years) |
| climate change | a long-term change in the weather patterns of a particular place |
| global warming | the rise in average temperature of the atmosphere near Earth's surface |
| mitigate | to lessen the harmful effects of something |
| weather | the outside air conditions of a specific place at a specific time, including variables such as wind speed, temperature, and humidity |

## TEACHER BACKGROUND INFORMATION
### Weather and Climate

An in-depth knowledge of meteorology is not necessary for this lesson. Do become familiar with the Climate Science Literacy principles at *https://downloads.globalchange. gov/Literacy/climate_literacy_highres_english.pdf.* According to these principles, a climate-literate person

- *understands the essential principles of Earth's climate system,*

- *knows how to assess scientifically credible information about climate,*

- *communicates about climate and climate change in a meaningful way, and*

- *is able to make informed and responsible decisions with regard to actions that may affect climate. (USGCRP 2009, p. 4)*

There are seven essential principles of climate science:

1. *The Sun is the primary source of energy for Earth's climate system.*

2. *Climate is regulated by complex interactions among components of the Earth system.*

3. *Life on Earth depends on, is shaped by, and affects climate.*

4. *Climate varies over space and time through both natural and man-made processes.*

5. *Our understanding of the climate system is improved through observations, theoretical studies, and modeling.*

6. *Human activities are impacting the climate system.*

7. *Climate change will have consequences for the Earth system and human lives. (USGCRP 2009, pp. 10–16)*

To understand climate change, you need to have a firm understanding of *weather, climate,* and the differences between the two. *Weather* is the current conditions of a specific place and time and includes aspects such as wind speed, temperature, and humidity as variables. *Climate* is an average of weather that takes place in a certain location. A main difference between the two is time. Weather takes place over short period of time, whereas climate is the average of daily weather for a certain location over an extended period of time. Another way to express the difference between weather and climate to students is to tell them, "Climate is what you expect, and weather is what you get." For example, you would expect to find a warm climate in the southwestern United States in May. However, you might need a sweater because of the weather, such as when a thunderstorm brings cooler temperatures. Extreme weather occurs when daily weather is significantly different from daily averages. Examples of extreme weather include floods, droughts, tornadoes, hurricanes, and snow and ice storms. Weather models can accurately predict weather for approximately 3 to 5 days in the future, while climate models can predict climate for a much longer time, such 30 years or more in the future (see *https://climatekids.nasa.gov/climate-model*).

## Layers of Earth's Atmosphere

Earth's atmosphere has five basic layers: troposphere, stratosphere, mesosphere, thermosphere, and exosphere. Weather takes place in the lower layers. Humans spend the majority of their time in the troposphere, but those on airplanes often journey into the stratosphere. Table 4.3 (p. 50) provides some attributes of each layer of Earth's atmosphere. More information can be found at *www.nesdis.noaa.gov/content/peeling-back-layers-atmosphere*.

**Table 4.3.** Attributes of Earth's Atmospheric Layers

| Part of Atmosphere | Height From Earth's Surface | Attributes |
|---|---|---|
| troposphere | 0–7 miles high (0–11 km) | • This is the first layer of the atmosphere, beginning at Earth's surface.<br>• This layer contains the air we breathe and most clouds.<br>• The air is warmer at the bottom because of heat from Earth. |
| stratosphere | 7–31 miles high (11–50 km) | • Ozone is formed here.<br>• The temperature rises with altitude. |
| mesosphere | 31–50 miles high (50–80 km) | • Meteors burn up on entry into this layer.<br>• The highest clouds form here.<br>• The average temperature is –120°F. |
| thermosphere | 50–440 miles high (80–708 km) | • This layer absorbs solar radiation.<br>• Temperatures range from –184°F to 3630°F. |
| exosphere | 440–6,200 miles high (708–9,979 km) | • This is the outermost layer, which touches outer space.<br>• The atmosphere is extremely thin here.<br>• Satellites orbit Earth in this layer. |

## Global Average Temperatures

The global temperature record is an average of surface temperatures across the planet. Global average temperatures are important for climate scientists to study because temperatures are vastly different at different locations around the world. The average global temperature has been steadily on the rise, as illustrated in Figure 4.1. A 1 degree increase in global temperature is significant because a great amount of energy is required to heat up all the ocean water, rivers, lakes, and land surfaces on the planet.

## Scientific Argumentation

The *Next Generation Science Standards* (NGSS Lead States 2013) and *A Framework for K–12 Science Education: Practices, Crosscutting Concepts, and Core Ideas* (NRC 2012) both emphasize the need to have students use scientific argumentation to support their knowledge claims in the classroom. Scientist use argumentation when they make claims that are based on observable evidence. NSTA Press's *Argument-Driven Inquiry* series provides

**Figure 4.1.** Change in Global Temperature Relative to 1951–1980 Average Temperatures

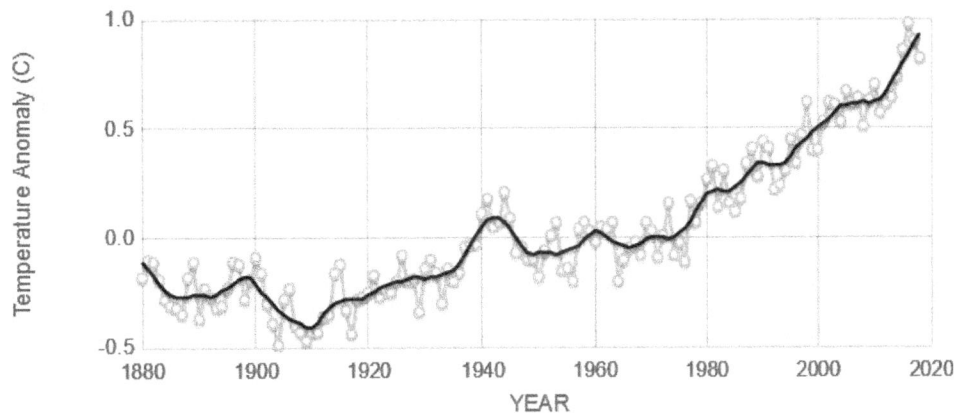

*Source: https://climate.nasa.gov/vital-signs/global-temperature.*

*Note:* Gray dots = annual mean; black line = lowess smoothing.

a format for a scientific argument that can be used to help students include all parts of their claim, evidence, and justification or reasoning. The graphic diagram of the structure of an argument is found in Figure 4.2.

For more information about scientific argumentation, see the following resources, both of which can be accessed for free online in PDF form:

- Practice 6, "Constructing Explanations and Designing Solutions," and Practice 7, "Engaging in Argument From Evidence," pages 61–74 of *A Framework for K–12 Science Education* (NRC 2012), available at

**Figure 4.2.** Framework of a Scientific Argument

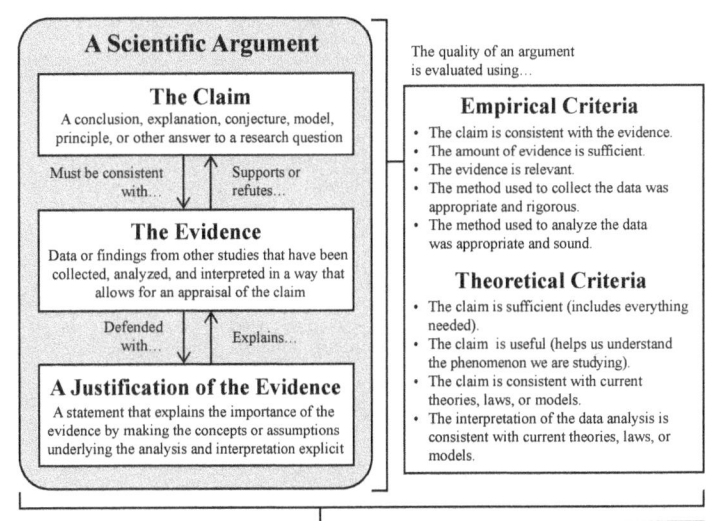

*Source:* Sampson, V., A. Murphy, K. Lipscomb, and T. L. Hutner. 2018. *Argument-driven inquiry in Earth and space science: Lab investigations for grades 6–10.* Arlington, VA: NSTA Press, p. 7.

*www.nap.edu/catalog/13165/a-framework-for-k-12-science-education-practices-crosscutting-concepts.*

- Chapter 5, "Generating and Evaluating Scientific Evidence and Explanations," pages 129–160 in *Taking Science to School* (NRC 2007), available at *www.nap.edu/catalog/11625/taking-science-to-school-learning-and-teaching-science-in-grades.*

## COMMON MISCONCEPTIONS

Students will have various types of prior knowledge about the concepts introduced in this lesson. Table 4.4 outlines some common misconceptions students may have concerning these concepts. Because of the breadth of students' experiences, it is not possible to anticipate every misconception that students may bring as they approach this lesson. Incorrect or inaccurate prior understanding of concepts can influence student learning in the future, however, so it is important to be alert to misconceptions such as those presented in the table.

**Table 4.4. Common Misconceptions About the Concepts in Lesson 1**

| Topic | Student Misconception | Explanation |
|---|---|---|
| Weather and climate | Weather and climate are the same. | Climate and weather are related but are not the same thing. Weather is variable and changes daily. It is the atmospheric conditions at a given place and time. Climate describes the average weather at a given place over a long period of time, at least 30 years. |
| Global warming | The world has been cooling for the past decade. | A decade is not a long enough period of time to explore overall trends. |
| | Scientists do not agree on the existence or causes of global climate change. | There is consensus among the majority of Earth scientists that the emission of greenhouse gases from human activities makes large contributions to current global warming trends. |
| | Water vapor in the atmosphere is the greenhouse gas primarily responsible for global warming. | There is an increase in water vapor in the atmosphere, but this is in response to the increased $CO_2$. |

## PREPARATION FOR LESSON 1

Review the Teacher Background Information section (p. 48) and preview the recommended videos in the Learning Components section to ensure that you have the necessary foundational knowledge about weather, climate, and climate change to teach this lesson. Assemble the materials for the lesson and make copies of student handouts. Prepare a sample Layers of Earth's Atmosphere Foldable. Have your students set up their STEM Research Notebooks (see pp. 24–25 for discussion and student instruction handout). Students should include all work for the module in the STEM Research Notebook, so you may wish to have them include section dividers in their notebooks.

## LEARNING COMPONENTS
### Introductory Activity/Engagement

**Connection to the Challenge:** After launching the module (see Module Launch, p. 26), begin each day of this lesson by directing students' attention to the driving question for the module challenge: How can we develop a local response to address an aspect of human impact on global climate change? This lesson helps scaffold students' foundational understanding of weather and climate; understanding these concepts is critical to understanding global warming and climate change.

**Pre-Assessment:** Have students complete the Climate Change Pre-Test (p. 63); collect the tests as students finish. Then, have students draw their environment using the Drawing of My Environment handout (p. 65). Tape their completed drawings on the wall for a gallery walk during this lesson.

**Science Class and Social Studies Connection:** Arrange students in small groups of three or four, and give each group several markers and a piece of butcher paper or large piece of paper. Hold a class discussion on students' responses to the first three questions on the pre-test. After a few minutes, ask students to think and talk about the role that humans play in climate change. Next, have each group create a poster that displays their best understanding of these terms and humans' role in climate change. Encourage students to add drawings to their poster to help explain their reasoning and understanding. Have student groups share their ideas with the class while you make a master list. Be sure to listen for any misconceptions about climate change and global warming (see Table 4.4 for some common misconceptions).

Next, invite students to go on a gallery walk of the environment drawings. As they walk around, ask students to note some of the similarities and differences among the pictures. Ask these facilitating questions to help students make connections to the main components of the environment:

- Do the pictures include *humans?*

- Do the pictures include *other living organisms* (e.g., plants or animals)?

- Do the pictures include *abiotic* items (e.g., mountains, rivers, the Sun, or clouds)?

- Do the pictures include *human built or designed* items (e.g., buildings, cars, or bridges)?

- Does each of these components appear to be interacting with other components in the environment?

It is very common for students not to include humans in their drawings of the environment. One of the goals of this unit is for students to learn to include themselves and other humans as part of the environment. Remind students that over the course of the next few weeks, they will learn about weather and climate, climate change, and how humans influence climate change and the environment. In addition, they are going to identify a local problem related to climate change and develop a solution to help mitigate that problem.

At the end of class, have students respond to the following on index cards and turn them in as exit tickets: *One question that I have about climate change is …?* After students have turned in their questions, prepare for the next class by compiling these questions into categories and creating a list of climate change questions that can be viewed by students at the start of each lesson. Leave space to add students' ideas about the answers to these questions throughout the module (for example, label pieces of chart paper with each of the questions, or create an electronic list that can be viewed by students).

**Mathematics Connection:** Using the U.S. Climate Data website (*www.usclimatedata.com*), have students explore the average temperature by month for various cities across the United States. Encourage them to explore cities in different regions. Figure 4.3 provides an example for one city from this website. Have students jot down in their STEM Research Notebooks what they notice about the shapes of the graphs, connections between the graphs for average low and high temperatures, and similarities and differences among different cities. Then, hold a class discussion on what they noticed. Tell them that the lines on the graph are based on the average temperature for that month at this location over all the years that

**Figure 4.3. Example From U.S. Climate Data Website**

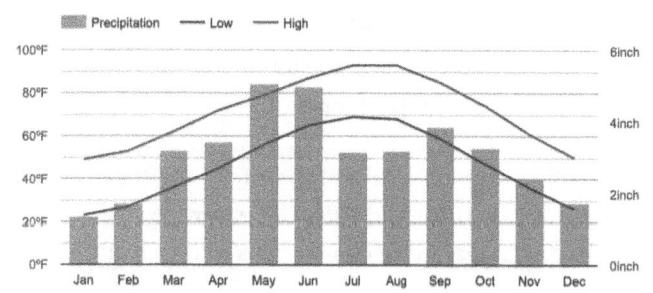

*Source:* U.S. Climate Data website (*www.usclimatedata.com/climate/ stillwater/oklahoma/united-states/usok0507*).

4

scientists have recorded climate data. (*Note:* This will help students later to make connections to the differences between weather and climate—you are planting a seed here.)

**ELA Connection:** Not applicable.

## Activity/Exploration

**Science Class:** Display a list of students' questions about climate change. Review the questions and hold a class discussion of how students might find answers to these questions. Tell students that they will return to these questions periodically throughout the module and attempt to answer them.

Students need a firm understanding of the differences between weather and climate before they can thoroughly understand climate change and its causes. Start a class discussion about weather. Discussion prompts might include the following:

- What was the weather like yesterday? Today?

- What will it be like tomorrow?

- Is the weather easy to predict?

Ask students to open their STEM Research Notebooks to the next new page and make a T-chart with the headings *Weather* and *Climate* at the top (see Figure 4.4).

**Figure 4.4.** Sample T-Chart for Comparing Weather and Climate

| Weather | Climate |
| --- | --- |
|  |  |

Show students a video that illustrates the basics of weather, and have students make notes of some of the things they learn from the video about weather in their T-charts. (*Note:* If students hear something about climate, they can note it in the chart as well, but they will revisit this T-chart for climate later.) Here are some example videos:

- "Weather 101," *https://video.nationalgeographic.com/video/101-videos/weather-101-sci*

- "Severe Weather: Crash Course Kids #28.2," *www.youtube.com/watch?v=QVZExLO0MWA*

- "Weather Channels: Crash Course Kids #34.2," *www.youtube.com/watch?v=RD-2dvaG4UY*

**Mathematics Connection:** Using the Weather Channel website (*https://weather.com*), have students choose "10 Day" from the menu bar, then enter their location (city, state) to see the local weather forecast (see Figure 4.5 as an example). Have students record the forecast for the next 10 days in their STEM Research Notebooks. Review the structure of an argument with students (see p. 51) and hand out the Claim, Evidence, Reasoning student handouts to each student (p. 66). Have each student make a claim (prediction), based on evidence and reasoning, about what they expect the high and low temperatures to be on the following day (day 6). Tell them to record their claim, evidence, and reasoning on the handouts. Ask students how these data are similar to or different from the data they explored on the U.S. Climate Data website.

**Figure 4.5.** Partial Sample 10-Day Weather Forecast for Stillwater, Oklahoma, From the Weather Channel Website

### Stillwater, OK 10 Day Weather
10:09 am CDT

| DAY | | DESCRIPTION | HIGH / LOW | PRECIP | WIND | HUMIDITY |
|---|---|---|---|---|---|---|
| **TODAY** APR 21 | | Partly Cloudy | 81°/58° | 0% | SE 10 mph | 37% |
| **WED** APR 22 | | Thunderstorms | 65°/53° | 100% | ESE 15 mph | 85% |
| **THU** APR 23 | | Sunny | 81°/57° | 10% | WNW 8 mph | 57% |
| **FRI** APR 24 | | Thunderstorms | 71°/48° | 80% | NNE 13 mph | 69% |
| **SAT** APR 25 | | Partly Cloudy | 71°/47° | 20% | NNW 15 mph | 53% |

*Source:* Search functions at *www.weather.com*.

*Note:* The full version of this weather forecast is available on the book's Extras page at *www.nsta.org/roadmap-humanimpacts*.

**ELA Connection:** Not applicable.

**Social Studies Connection:** Not applicable.

## Explanation

**Science Class:** Hold a class discussion on what students learned from the video about weather that they watched the previous day. Possible discussion prompts could include the following:

- What do you notice about weather patterns?

- Is weather the same throughout the year?

- How predictable is the weather?

- Do we expect to have the same kinds of weather every day in our town? How so?

Help students conclude that weather is what happens in our atmosphere on a daily basis. It varies based on factors such as wind patterns, air moisture content, and the time of year. Tell students that to better understand weather and climate, they will watch a video to help them understand the layers of Earth's atmosphere. Here are some example videos:

- "Earth's Atmosphere," *http://studyjams.scholastic.com/studyjams/jams/science/ weather-and-climate/earths-atmosphere.htm*

- "Layers of Atmosphere: The Dr. Binocs Show," *www.youtube.com/ watch?v=5sg9sCOXFIk*

### STEM Research Notebook Prompt

Have students respond to the following questions in their STEM Research Notebooks as they watch the video: *What are the various layers in the sky? What takes place in each layer?*

Then, have students use their notes about the different layers of the atmosphere to help them construct a Layers of Earth's Atmosphere Foldable (pp. 67–71) to put into their notebooks.

**Mathematics Connection:** Not applicable.

**ELA and Social Studies Connections:** Have students do internet research to explore the concept of invention. They should examine the history of human invention and how major inventions transformed society, including both positive and negative effects. Various published timelines of invention are available to support this inquiry, including the following:

- *www.britannica.com/story/history-of-technology-timeline*

- *www.history.com/news/11-innovations-that-changed-history*

- *www.visualcapitalist.com/wp-content/uploads/2015/04/worlds-greatest-inventions.html*

## Elaboration/Application of Knowledge

**Science Class:** Tell students they are going to analyze actual data of changes in average temperatures measured on Earth for more than 139 years. Students will examine data from the National Oceanic and Atmospheric Administration (NOAA) found at *https:// climate.nasa.gov/vital-signs/global-temperature/*. By hovering the cursor over each point on the graph, they can see the average anomaly (change) in temperature (in Celsius) for each year from 1880 to 2019, meaning how much the average global temperature differed from the 1951–1980 average.

Divide students into groups of two or three. We recommend giving each group of students a decade's worth of data points to plot on their graph, but all data points should be plotted, so you may need to assign some groups to plot more than one set of data. Each group should plot the years on the *x*-axis and the changes in temperature on the *y*-axis. It is important that all groups use the same scale for their horizontal and vertical axes so they can combine their graphs into one larger graph. Thus, you can either have students work together to decide what the scale should be, or you can provide graph paper. Tell students to make large, bold marks so their graph lines will be visible for all to see when displayed on the wall.

Once groups are done plotting their data, they should reflect on the data for their assigned decade. Do they see any trends in the changes in annual temperature for their decade? Next, have students form larger groups to examine trends for the 1800s, 1900–1949, 1950–1999, and the 2000s. What trends do students notice in these larger time frames?

Finally, have students tape all their graphs together and hang them on a wall so everyone can see them. (If you don't have enough wall space in your room, you may be able to display the graphs in the hallway.) Have students examine the entire timeline for trends from 1880 to present. Lead a discussion with students about how it is hard to see trends in temperature data with only a small data set. As scientists collect more data, they can become more aware of trends.

Ask students to sketch the data trend of changes in temperature over time in their STEM Research Notebooks. Then, have students use the claim, evidence, reasoning model to summarize the trend of data from the past century in their notebooks.

Next, have students watch a video on the differences between weather and climate, such as "Weather Versus Climate Change" at *www.youtube.com/watch?v=cBdxDFpDp_k* or "What Is Climate?" at *https://planetnutshell.com/portfolio/what-is-climate*.

Ask students to revisit their T-charts on the differences between weather and climate. Have them add to their definitions, making sure that they understand the following:

- *Weather* is the current conditions of a specific place and time and includes aspects such as wind speed, temperature, and humidity as variables.

- *Climate* is the average weather conditions over a long period of time, typically no less than 30 years.

## STEM Research Notebook Prompt

Have students respond to the following questions in their STEM Research Notebooks:

- *Over the past century, has Earth been cooling, warming, or staying the same? Provide evidence for your response.*

- *What are the differences between weather and climate?*

**Mathematics Connection:** Again using the Weather Channel website (*https://weather. com*), students should now look at the monthly forecast. Ask them to examine how the daily highs and lows compare with the historical average listed for that month. Then, have students calculate the mean (arithmetic average) for the month based on the forecast temperatures and see how this mean compares with the historical monthly average. Hold a class discussion asking questions such as the following: How do the forecast daily temperatures compare with the historical monthly average? How would a rare cold or hot front (depending on the month) influence the average monthly temperature for that year? How would those data then influence the historical monthly average temperature?

**ELA Connection:** Have students do internet research to explore the concept of engineering and an engineering design process. Ask students to use their research and the research they conducted about inventors to answer the following questions in a class discussion:

- What is engineering?

- How is the work that engineers and inventors do similar?

- How is the work that engineers and inventors do different?

- How do engineers use science and mathematics in their work?

- How do inventors use science and mathematics in their work?

Have each student choose an object they think was created by an inventor or engineer and do research to learn about the person who created that object. Have students answer the following questions in their STEM Research Notebooks:

- What object are you researching?

- Why are you interested in this object?

- Who first created this object, and when?

- Why did this person create this object?

- Did this person invent or design this object as part of their job?

- What kinds of skills and knowledge did the person use to create this object?

- What kind of process did the person use to create this object?

- How did the creation of this object influence people who used it?

- Have other people made improvements to this object over time? If so, what kinds of improvements?

**Social Studies Connection**: Have students participate in simulations of data use. In the first simulation, have students vote on a rule for the classroom that is illogical and will cause some brief chaos, such as that students may sharpen their pencils only when one of their classmates is sharing. Provide data that are intentionally biased and based on only one side of an argument so that you can address how data can be used in misleading ways (e.g., you might state that most students prefer to sharpen their pencils when their peers are sharing and that there are benefits to getting exercise while learning). Invite some students to play the role of "experts" to share why the rule is important based on the biased data. Then, have the class vote on the rule. Have the rule enacted for a portion of social studies class. After it causes some chaos for a brief moment, stop the class and engage students in a discussion. What made some of them vote for the rule? What could have been a better way to ensure that the data were accurate and that both sides of the argument were heard? Engage them in a discussion about the importance of being an informed citizen. Have students develop their own simulation in which they come up with a rule that is fair and then enact it.

Discuss with students that global warming and climate change are often topics in the news and in debates. What can they do to ensure that they are getting accurate and unbiased information about global warming and climate change?

## Evaluation/Assessment

Students may be assessed on the following performance tasks and other measures listed.

*Performance Tasks*

- Graph of changes in average global temperature (p. 51)

- Climate Change Pre-Test (p. 63)

- Environment drawing (p. 65)

- Claim, Evidence, Reasoning Student Handout (p. 66)

*Other Measures*

- STEM Research Notebook Entry Rubric (p. 72)

- Collaboration Rubric (p. 73)

- Participation in class discussions

## INTERNET RESOURCES

Climate Science Literacy principles
- *https://downloads.globalchange.gov/Literacy/climate_literacy_highres_english.pdf*

Climate models
- *https://climatekids.nasa.gov/climate-model*

Layers of Earth's atmosphere
- *www.nesdis.noaa.gov/content/peeling-back-layers-atmosphere*

U.S. Climate Data
- *www.usclimatedata.com*

"Weather 101" video
- *https://video.nationalgeographic.com/video/101-videos/weather-101-sci*

"Severe Weather: Crash Course Kids #28.2" video
- *www.youtube.com/watch?v=QVZExLO0MWA*

"Weather Channels: Crash Course Kids #34.2" video
- *www.youtube.com/watch?v=RD-2dvaG4UY*

The Weather Channel
- *https://weather.com*

"Earth's Atmosphere"
- *http://studyjams.scholastic.com/studyjams/jams/science/weather-and-climate/earths-atmosphere.htm*

"Layers of Atmosphere: The Dr. Binocs Show" video
- *www.youtube.com/watch?v=5sg9sCOXFIk*

Invention timelines
- *www.britannica.com/story/history-of-technology-timeline*

- *www.history.com/news/11-innovations-that-changed-history*

- *www.visualcapitalist.com/wp-content/uploads/2015/04/worlds-greatest-inventions.html*

Changes in average global temperature
- *https://climate.nasa.gov/vital-signs/global-temperature*

"Weather Versus Climate Change: Cosmos: A Spacetime Odyssey" video
- *www.youtube.com/watch?v=cBdxDFpDp_k*

"What Is Climate?"
- *planetnutshell.com/portfolio/what-is-climate*

4

Name: _____     Date: _____

# CLIMATE CHANGE PRE-TEST

1. What is climate?

2. What is weather?

3. What do you know about climate change?

4. How well do you feel you understand the issue of climate change? **(circle one)**

   Very well        Fairly well        Not very well        Not at all        No opinion

5. How well do you feel you understand the issue of global warming? **(circle one)**

   Very well        Fairly well        Not very well        Not at all        No opinion

6. Do you believe increases in Earth's temperature over the last century are due more to … **(select one)**

   a. Human activities

   b. Natural causes

   c. Don't know

Name: _____          Date: _____

**STUDENT HANDOUT, PAGE 2**

# CLIMATE CHANGE PRE-TEST

7. In your opinion, when will the effects of climate change begin to happen? **(select one)**

    a. They will never happen.

    b. They will not happen in my lifetime, but future generations will feel the effects.

    c. They will start happening within my lifetime.

    d. They will start happening in the next few years.

    e. They are happening now.

True/False **(circle one)**

8. Weather and climate are the same thing.                                          True   or   False

9. Global warming and the greenhouse effect are the same thing.                     True   or   False

10. Climate change and global warming are the same thing.                           True   or   False

11. Carbon dioxide is a greenhouse gas.                                             True   or   False

12. Greenhouse gases capture heat and warm Earth and its atmosphere.                True   or   False

13. Greenhouse gases are caused only by humans.                                     True   or   False

14. Scientists agree about whether climate change is happening.                     True   or   False

15. Climate change is happening as a result of an increase in
    greenhouse gases in the atmosphere.                                            True   or   False

16. There is nothing that I can do to lessen the effects of climate change.         True   or   False

Name: _____

Date: _____

**STUDENT HANDOUT**

# DRAWING OF MY ENVIRONMENT

In the space below, draw a picture of your environment.

*Source:* Adapted from Moseley, Perrotta, and Utley (2010).

**STUDENT HANDOUT**

## CLAIM, EVIDENCE, REASONING

| Problem or Question |
| --- |
| |

| Claim |
| --- |
| (This is your prediction of the solution to the problem or answer to the question.) |
| |

| Evidence | Reasoning |
| --- | --- |
| (What data are you using to support your claim?) | (How do you justify using this evidence to support your claim? Why does this make sense?) |
| | |

Name: _____     Date: _____

**STUDENT HANDOUT, PAGE 1**

## LAYERS OF EARTH'S ATMOSPHERE FOLDABLE

### DIRECTIONS

*Note:* Each set of circles makes two foldables.

1. Cut out the circles.

2. Cut each circle in half along the dotted lines.

3. Lay the largest half circle on your desk. Place the next smallest on top of it, about 1/8 inch from the straight edge of the largest half circle (this creates a lip that you will fold up over the other pieces to secure all the half circles together).

4. Stack the other half-circles by decreasing size, aligning the straight edges of each (the smallest with Earth should be on top).

5. Fold up the 1/8 inch exposed lip of the largest half circle to cover the straight edges of the other pieces, and staple the folded edge at the edges and in the middle.

6. Label the outside edge of each circle (from smallest to largest): Troposphere, Stratosphere, Mesosphere, Thermosphere, and Exosphere

7. Fold down the different spheres and record different facts about each sphere on the back of the half-circle (see table that follows).

Name: _____     Date: _____

**STUDENT HANDOUT, PAGE 2**

## LAYERS OF EARTH'S ATMOSPHERE FOLDABLE

| Part of Atmosphere | Height From Earth's Surface | Attributes |
|---|---|---|
| Troposphere | 0–7 miles high (0–11 km) | First layer of the atmosphere that contains most of the atmosphere's mass. |
| Stratosphere | 7–31 miles high (11–50 km) | Ozone is formed here. The temperature rises with increasing altitude. |
| Mesosphere | 31–50 miles high (50–80 km) | Meteors burn up upon entry. The highest clouds form here. The average temperature is –120°F. |
| Thermosphere | 50–440 miles high (80–708 km) | Absorbs solar radiation. Temperature ranges from –184°F to 3630°F but would feel cold to us because the atmosphere is very thin. |
| Exosphere | 440–6,200 miles high (708–9,979 km) | Outermost layer that touches outer space. Atmosphere is extremely thin. Satellites are placed here. |

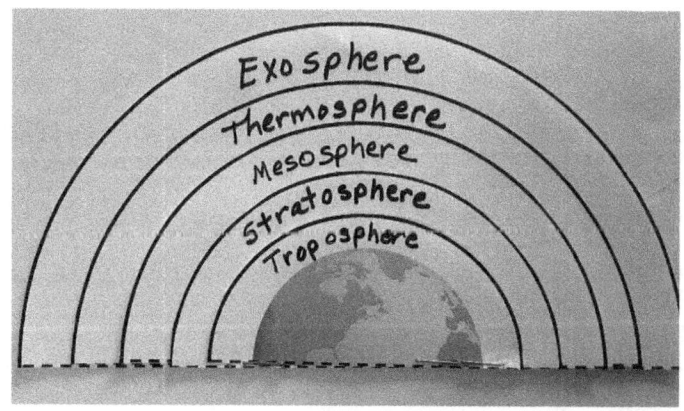

NATIONAL SCIENCE TEACHING ASSOCIATION

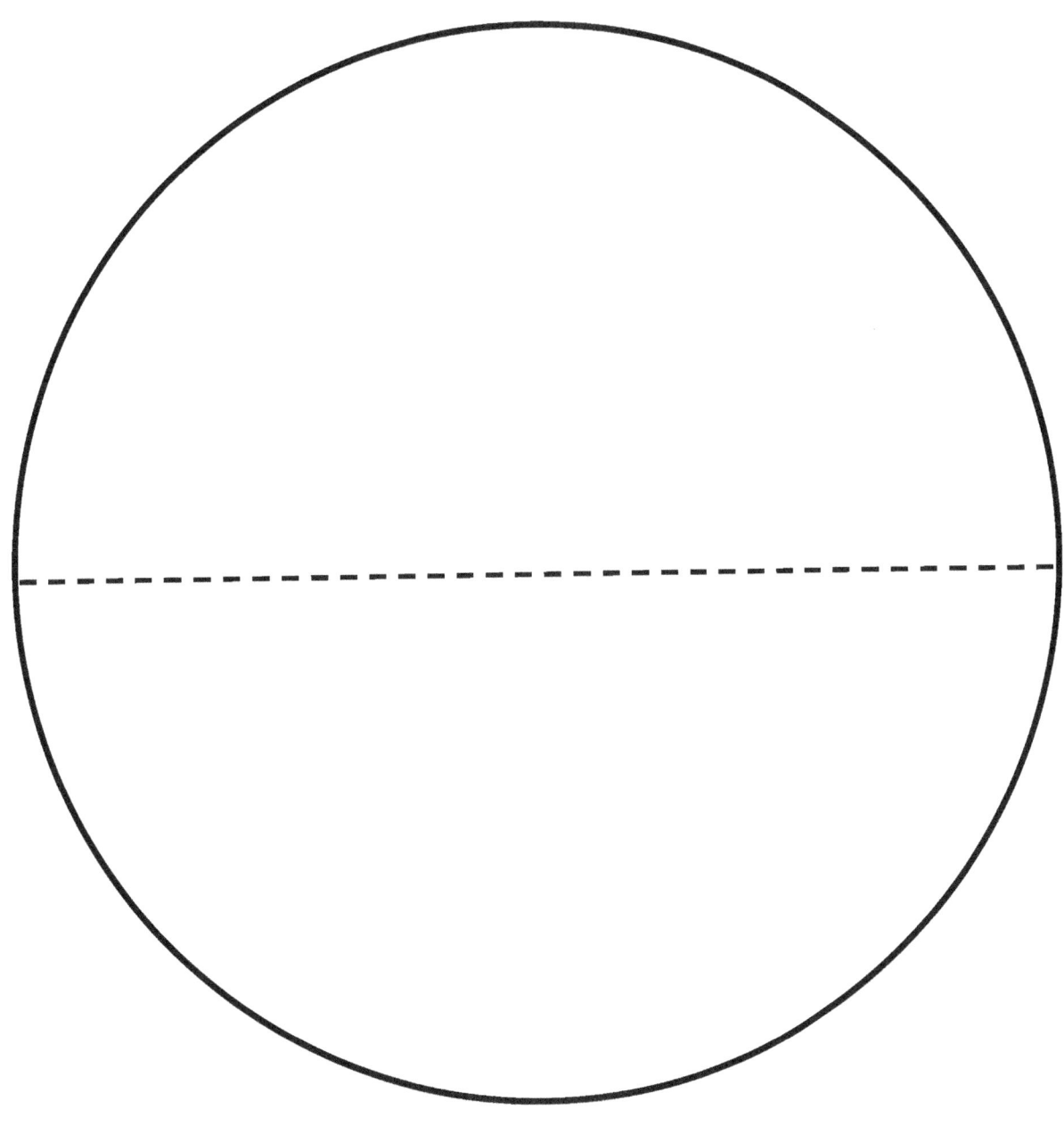

Human Impacts on Our Climate Lesson Plans

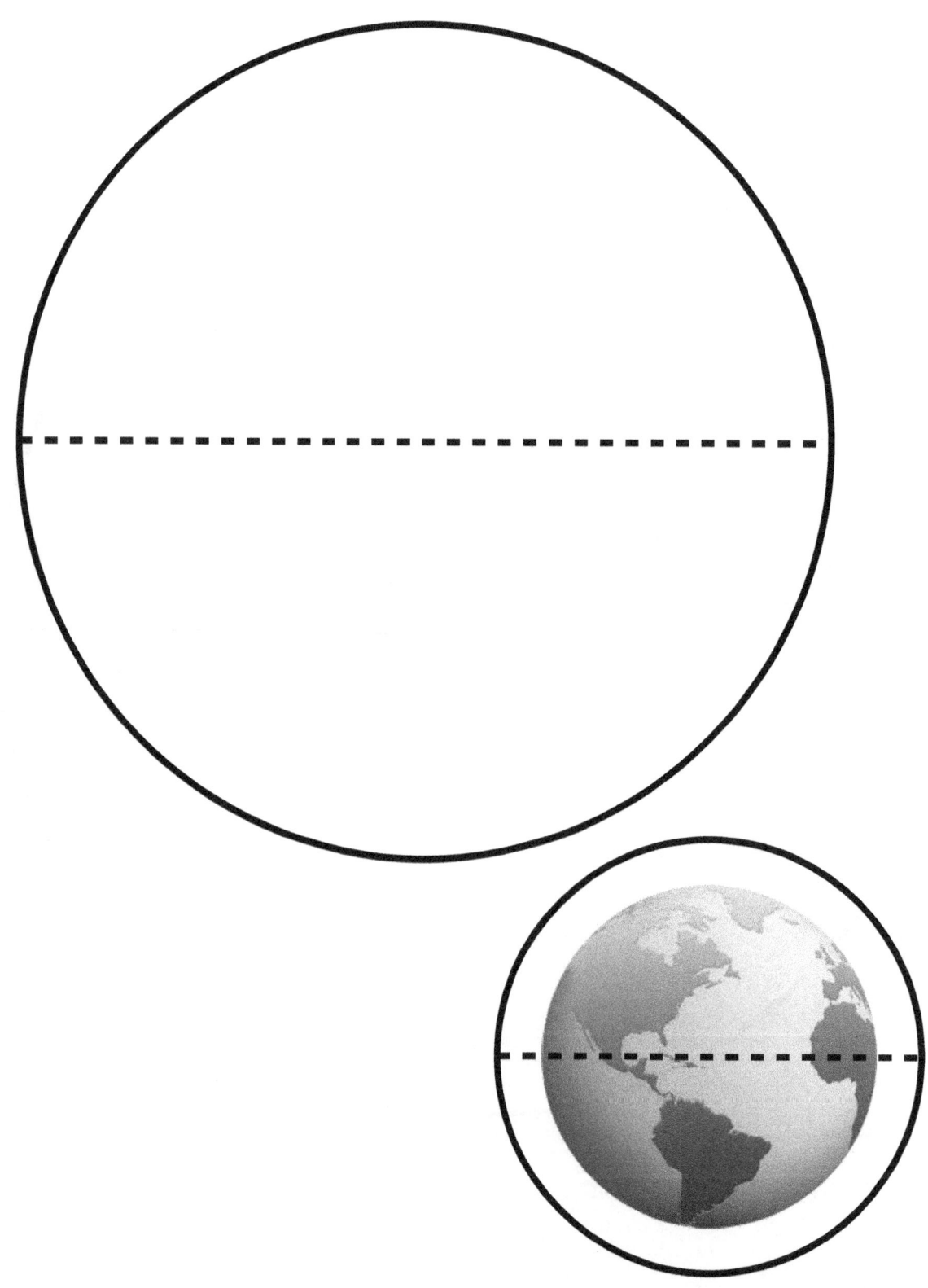

NATIONAL SCIENCE TEACHING ASSOCIATION

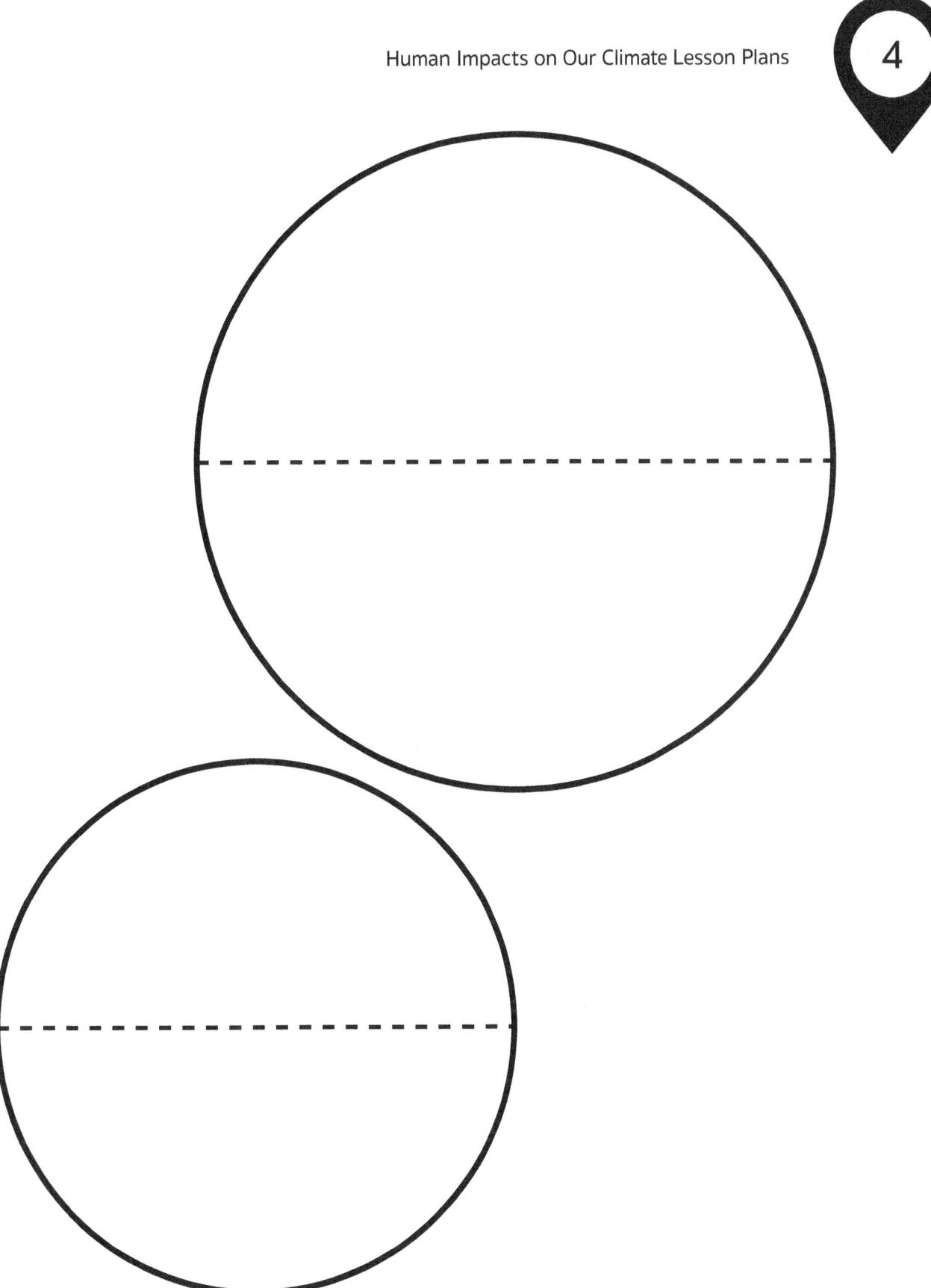

| | STEM Research Notebook Entry Rubric | | | |
|---|---|---|---|---|

Name: _____

| Criteria | Beginning/Does Not Meet Minimum Expectations (1 point) | Progressing/Does Not Fully Meet Expectations (2 points) | Competent/Meets Expectations (3 points) | Score |
|---|---|---|---|---|
| CLAIM | No claim is made or the claim is inaccurate. | An accurate claim is made, but it is vague or incomplete. | A complete and accurate claim is made. | |
| EVIDENCE | Evidence is not provided or the evidence does not support the claim. | Appropriate evidence is provided, but it is insufficient to support the claim. May also provide some inappropriate evidence. | Evidence provided sufficiently supports claim. | |
| REASONING | No reasoning is provided or reasoning is provided but is not linked to evidence or claim. | Reasoning repeats evidence but does not sufficiently link it to some of the scientific principles. | Accurate and complete reasoning that links the claim and evidence is provided. It is sufficiently supported by scientific principles. | |

TOTAL SCORE: _____

COMMENTS:

## Collaboration Rubric

Name: _____

| Skills | Beginning/Does Not Meet Minimum Expectations (1 point) | Progressing/Does Not Fully Meet Expectations (2 points) | Competent/Meets Expectations (3 points) | Advanced (4 points) | Score |
|---|---|---|---|---|---|
| CONTRIBUTION, PARTICIPATION, ATTITUDE | Seldom cooperative. Rarely offers useful ideas. Is disruptive. | Sometimes cooperative. Sometimes offers useful ideas. Rarely displays positive attitude. | Cooperative. Usually offers useful ideas. Generally displays positive attitude. | Always willing to help and do more. Routinely offers useful ideas. Always displays positive attitude. | |
| WORKING WITH OTHERS, COOPERATION | Does not contribute or do any work. Does not work well with others. Often argues with teammates. | Could have done more of the work. Has difficulty and requires extra structure, directions, and leadership. Sometimes argues. | Does his or her share of the work. Is cooperative. Works well with others. Rarely argues. | Does more than others. Is highly productive. Works extremely well with others. Never argues. | |
| FOCUS ON TASK, COMMITMENT | Often is not a good team member. Does not focus on the task and what needs to be done. Lets others do the work. | Sometimes is not a good team member. Sometimes focuses on the task and what needs to be done. Must be prodded and reminded to keep on task. | Does not cause problems in the group. Focuses on the task and what needs to be done most of the time. Can be counted on. | Tries to keep people working together. Almost always is focused on the task and what needs to be done. Is very self-directed. | |
| COMMUNICATION, LISTENING, INFORMATION SHARING | Rarely listens to, shares with, or supports the efforts of others. Is always talking and never listens to others. Provides no feedback to others. Does not relay any information to teammates. | Sometimes listens to, shares with, and supports the efforts of others. Usually does most of the talking and rarely listens to others. Provides little feedback to others. Relays very little information, some related to the topic. | Usually listens to, shares with, and supports the efforts of others. Sometimes talks too much. Provides some effective feedback to others. Relays some basic information, most related to the topic. | Always listens to, shares with, and supports the efforts of others. Provides effective feedback to other members. Relays a great deal of information, all related to the topic. | |

TOTAL SCORE: _____

COMMENTS:

# Lesson Plan 2: The Greenhouse Effect and Climate Change

In this lesson, students make models of greenhouses to study the greenhouse effect. Then, they explore a computer simulation to understand greenhouse gases and their connection to global temperatures and learn the difference between greenhouse gases and climate change. Finally, students study different climate change indicators.

## ESSENTIAL QUESTIONS

- What are greenhouse gases?

- What is the greenhouse effect?

- What role do greenhouse gases play in global temperature warming?

- How do greenhouse gases affect the climate?

## ESTABLISHED GOALS AND OBJECTIVES

At the conclusion of this lesson, students will be able to do the following:

- Explain which gases are greenhouse gases

- Know whether all greenhouse gases contribute to the greenhouse effect

- Collect and analyze data to study the greenhouse effect and to determine which gases are greenhouse gases

- Determine whether all greenhouse gases contribute to the greenhouse effect

- Make conclusions about the role of greenhouse gases in Earth's atmosphere, consider how greenhouse gases affect our climate, and predict what will happen to average global temperature

- Use scientific data to develop an informed position on changes in Earth's climate

- Identify potential causes of climate change

- Identify how humans can influence climate change

- Describe concentrations of greenhouse gases during the Ice Age, preindustrial period, and present

- Describe average global temperature trends and greenhouse gas concentration levels over time

## TIME REQUIRED

- 6 days (approximately 45 minutes each day; see Tables 3.6 and 3.7, pp. 37–38)

## MATERIALS

*Required Materials for Lesson 2*

- STEM Research Notebooks

- Computer and projector for students to watch videos

- Computers, tablets, or laptops with internet access for student research and simulation (at least 1 per pair of students)

- Handouts (attached at the end of this lesson)

- 3" × 5" index cards (1 per student)

*Additional Materials for Greenhouse Models* (per team unless otherwise noted)

- 2 clear plastic 2-liter bottles

- Marker to label bottles

- Water (about 150 mL per bottle)

- Clay

- 2 thermometers (preferably digital probe type)

- 4 sodium bicarbonate tablets

- 2 small squares plastic wrap

- 2 large rubber bands

- Heat lamp

- Indirectly vented chemical splash goggles (1 pair per student)

- Nonlatex apron (1 per student)

## SAFETY NOTES

1. All students must wear safety goggles and nonlatex aprons during the setup, hands-on, and takedown segments of the activity.

2. Use caution when working with glass or plasticware, which can break and cut skin.

3. Immediately pick up any items dropped on the floor to avoid a slip-and-fall hazard.

4. Immediately wipe up any spilled water on the floor to avoid a slip-and-fall hazard.

5. Use caution when working with the heat lamp, which can burn skin on medium and high settings.

6. Use only GFI-protected circuits when using electrical equipment, and keep electrical wires away from water sources to avoid risk of shock.

7. Caution students not to place the tablets used in the lab activity in the mouth or swallow them.

8. Wash hands with soap and water after completing this activity.

## CONTENT STANDARDS AND KEY VOCABULARY

Table 4.5 lists the content standards from the *NGSS, CCSS,* and the Framework for 21st Century Learning that this lesson addresses, and Table 4.6 (p. 80) presents the key vocabulary. Vocabulary terms are provided for both teacher and student use. Teachers may choose to introduce some or all of the terms to students.

**Table 4.5.** Content Standards Addressed in STEM Road Map Module Lesson 2

---

*NEXT GENERATION SCIENCE STANDARDS*

**PERFORMANCE EXPECTATIONS**

- MS-ESS3-4. Construct an argument supported by evidence for how increases in human population and per-capita consumption of natural resources impact Earth's systems.

- MS-ESS3-5. Ask questions to clarify evidence of the factors that have caused the rise in global temperatures over the past century.

**SCIENCE AND ENGINEERING PRACTICES**

*Asking Questions and Defining Problems*

Asking questions and defining problems in grades 6–8 builds on grades K–5 experiences and progresses to specifying relationships between variables, and clarifying arguments and models.
- Ask questions to identify and clarify evidence of an argument.

---

*Continued*

**Table 4.5.** (*continued*)

*Analyzing and Interpreting Data*

Analyzing data in 6–8 builds on K–5 and progresses to extending quantitative analysis to investigations, distinguishing between correlation and causation, and basic statistical techniques of data and error analysis.

- Analyze and interpret data to determine similarities and differences in findings.

*Constructing Explanations and Designing Solutions*

Constructing explanations and designing solutions in 6–8 builds on K–5 experiences and progresses to include constructing explanations and designing solutions supported by multiple sources of evidence consistent with scientific ideas, principles, and theories.

- Apply scientific principles to design an object, tool, process or system.

*Engaging in Argument From Evidence*

Engaging in argument from evidence in 6–8 builds on K–5 experiences and progresses to constructing a convincing argument that supports or refutes claims for either explanations or solutions about the natural and designed world(s).

- Construct an oral and written argument supported by empirical evidence and scientific reasoning to support or refute an explanation or a model for a phenomenon or a solution to a problem.

DISCIPLINARY CORE IDEAS

*ESS3.C. Human Impacts on Earth Systems*

- Human activities have significantly altered the biosphere, sometimes damaging or destroying natural habitats and causing the extinction of other species. But changes to Earth's environments can have different impacts (negative and positive) for different living things.

- Typically as human populations and per-capita consumption of natural resources increase, so do the negative impacts on Earth, unless the activities and technologies involved are engineered otherwise.

CROSSCUTTING CONCEPTS

*Patterns*

- Graphs, charts, and images can be used to identify patterns in data.

*Cause and Effect*

- Relationships can be classified as causal or correlational, and correlation does not necessarily imply causation.

- Cause and effect relationship may be used to predict phenomena in natural or designed systems.

*Continued*

**Table 4.5.** (*continued*)

*Stability and Change*
- Stability might be disturbed either by sudden events or gradual changes that accumulate over time.

### COMMON CORE STATE STANDARDS FOR MATHEMATICS

#### MATHEMATICAL PRACTICES
- MP1. Make sense of problems and persevere in solving them.
- MP2. Reason abstractly and quantitatively.
- MP3. Construct viable arguments and critique the reasoning of others.
- MP4. Model with mathematics.
- MP5. Use appropriate tools strategically.
- MP6. Attend to precision.
- MP8. Look for and express regularity in repeated reasoning.

#### MATHEMATICAL CONTENT
- 6.SP.B.4. Display numerical data in plots on a number line, including dot plots, histograms, and box plots.
- 6.SP.5. Summarize numerical data sets in relation to their context.

### COMMON CORE STATE STANDARDS FOR ENGLISH LANGUAGE ARTS

#### READING STANDARDS
- RI.6.1. Cite textual evidence to support analysis of what the text says explicitly as well as inferences drawn from the text.
- RI.6.2. Determine a central idea of a text and how it is conveyed through particular details; provide a summary of the text distinct from personal opinions or judgments.
- RI.6.4. Determine the meaning of words and phrases as they are used in a text, including figurative, connotative, and technical meanings.
- RI.6.7. Integrate information presented in different media or formats (e.g., visually, quantitatively) as well as in words to develop a coherent understanding of a topic or issue.

#### WRITING STANDARDS
- W.6.1. Write arguments to support claims with clear reasons and relevant evidence.
- W.6.2. Write informative/explanatory texts to examine a topic and convey ideas, concepts, and information through the selection, organization, and analysis of relevant content.
- W.6.2A. Introduce a topic, organize ideas, concepts and information, using strategies such as definition, classification, comparison/contrast, and cause/effect; include formatting (e.g., headings), graphics (e.g. charts, tables), and multimedia when useful to aiding comprehension.

*Continued*

**Table 4.5.** (*continued*)

- W.6.2B. Develop the topic with relevant facts, definitions, concrete details, quotations, or other information and examples.

- W.6.2D. Use precise language and domain-specific vocabulary to inform about or explain the topic.

- W.6.2F. Provide a concluding statement or section that follows from the information or explanation presented.

- W.6.7. Conduct short research projects to answer a question, drawing on several sources and refocusing the inquiry when appropriate.

- W.6.8. Gather relevant information from multiple print and digital sources; assess the credibility of each source; and quote or paraphrase the data and conclusions of others while avoiding plagiarism and providing basic bibliographic information for sources.

**SPEAKING AND LISTENING STANDARDS**

- SL.6.1. Engage effectively in a range of collaborative discussions (one-on-one, in groups, and teacher-led) with diverse partners on grade 6 topics, texts, and issues, building on others' ideas and expressing their own clearly.

- SL.6.1.A. Come to discussions prepared, having read or studied required material; explicitly draw on that preparation by referring to evidence on the topic, text, or issue to probe and reflect on ideas under discussion.

- SL.6.1.B. Follow rules for collegial discussions, set specific goals and deadlines, and define individual roles as needed.

- SL.6.2. Interpret information presented in diverse media and formats (e.g., visually, quantitatively, orally) and explain how it contributes to a topic, text, or issue under study.

- SL.6.4. Present claims and findings, sequencing ideas logically and using pertinent descriptions, facts, and details to accentuate main ideas or themes; use appropriate eye contact, adequate volume, and clear pronunciation.

- SL.6.5. Include multimedia components (e.g., graphics, images, music, sound) and visual displays in presentations to clarity information.

**FRAMEWORK FOR 21ST CENTURY LEARNING**

- Interdisciplinary Themes: Global Awareness, Environmental Literacy

- Learning and Innovation Skills: Critical Thinking and Problem Solving, Communication, Collaboration

- Information, Media, and Technology Skills: Information Literacy

- Life and Career Skills: Productivity and Accountability

**Table 4.6. Key Vocabulary for Lesson 2**

| Key Vocabulary | Definition |
|---|---|
| anthropogenic | influences on nature that originate in human activity |
| climate change indicators | variables that scientists study to understand how Earth's climate is changing over time, such as global temperatures, glacial melt, precipitation, and animal migratory patterns |
| electromagnetic radiation | waves of energy and light that travel through the air around us and are made up of photons |
| greenhouse effect | a process in which gases in the atmosphere trap the Sun's heat and cause Earth's surface and troposphere to become warmer |
| greenhouse gases | gases that absorb infrared radiation, trap heat (energy) from the Sun in the atmosphere, and contribute to the greenhouse effect; examples are carbon dioxide and methane |
| infrared | describes a wave that is similar to a light wave but that the human eye cannot see; we can feel these waves as heat |
| photon | a particle of light or other electromagnetic radiation |
| sea level | the place where the ocean usually meets the land; used as the zero point to measure other land heights |
| tide | the rising and falling levels of ocean water caused by gravity from the Sun and Moon and by the Earth's rotation |

## TEACHER BACKGROUND INFORMATION
### Greenhouse Effect and Greenhouse Gases

The term *greenhouse effect* reflects the fact that Earth's lower atmosphere acts like a greenhouse, trapping heat from the Sun that has been reflected by Earth. The greenhouse effect is caused by a variety of *greenhouse gases,* including water vapor, carbon dioxide ($CO_2$), methane, nitrous oxide, ozone, and chlorofluorocarbons, which absorb heat reflected from Earth's surface. This is a natural process that is important to life on Earth. Much in the same way a greenhouse keeps plants warm in the winter, greenhouse gases keep Earth warm as they trap heat (i.e., radiation) in the atmosphere. Earth is a warm planet because it has an atmosphere that contains greenhouse gases. If there were no atmosphere, the radiation from the Sun would reflect back into space and Earth would be a very cold planet. Without the greenhouse effect, the planet would be covered in ice and too cold to support life. Historically, there have been variable concentrations of greenhouse gases in the atmosphere. However, since the Industrial Revolution, human

technological development has contributed to a major increase of greenhouse gases, primarily carbon dioxide, into the atmosphere.

Carbon dioxide is an important greenhouse gas. It is released in the atmosphere through natural processes such as volcanic eruptions and respiration, but it is also released through many human activities, including burning fossil fuels and deforestation. The Intergovernmental Panel on Climate Change (IPCC 2014), which is composed of over 1,300 independent scientific experts, concluded that there is more than a 90% probability that human activities over the past 250 years have resulted in increased temperatures on the planet. Because of the increase of carbon dioxide in the atmosphere from human activities, we now have an enhanced greenhouse effect. The increase in atmospheric carbon dioxide is increasing average global temperatures and is thus the primary driver of recent global climate change.

Figure 4.6 displays atmospheric carbon dioxide levels for the past 800,000 years. As illustrated by the dashed line in the figure, atmospheric carbon dioxide levels stayed below the level of around 300 parts per million (ppm) for hundreds of thousands of years. However, the level started to climb over the past century as a result of industrialization, and atmospheric carbon dioxide surpassed 400 ppm for the first time in 2013.

**Figure 4.6. Carbon Dioxide Levels for Past 800,000 Years**

![Graph showing carbon dioxide level (parts per million) on the y-axis ranging from 160 to 480, and years before today (0 = 1950) on the x-axis ranging from 800,000 to 0. A dashed line marks the level "For millennia, atmospheric carbon dioxide had never been above this line." Annotations point to "current level" and "1950 level."]

*Source:* NASA Global Climate Change: Vital Signs of the Planet. *https://climate.nasa.gov/climate_resources/24/graphic-the-relentless-rise-of-carbon-dioxide.*

*Note:* A full-color version of this figure is available on the book's Extras page at *www.nsta.org/roadmap-humanimpacts.*

## Climate Change and Climate Change Indicators

"Changing Planet: Past, Present, Future" is a lecture series at *www.hhmi.org/biointeractive/changing-planet-past-present-future* that summarizes some of what scientists know about climate change. In particular, view "Climate Change: How Do We Know We're Not Wrong?" at *www.hhmi.org/biointeractive/climate-change-how-do-we-know-were-not-wrong*.

Scientists study a variety of climate change indicators, measures of different variables that help them understand the causes of climate change and its impacts on Earth, such as changes in greenhouse gases, weather and climate, oceans, snow and ice, health and society, and ecosystems. You should become familiar with common climate change indicators, which include atmospheric concentrations of greenhouse gases, average global temperature, drought, flooding, ocean heat, sea surface temperature, ocean acidity, and precipitation rates. More information on climate change indicators can be found on the Environmental Protection Agency (EPA) website at *www.epa.gov/climate-indicators*. In addition to taking direct measurements of different indicators, scientists can use satellite images to study climate change. See NASA's "Global Climate Change: Vital Signs of the Planet" at *http://climate.nasa.gov* and "Images of Change" at *https://climate.nasa.gov/images-of-change*.

Human activities have altered the greenhouse effect that naturally exists on Earth. Burning fossil fuels (coal, oil, and natural gas) has increased the concentrations of carbon dioxide in the atmosphere. Scientists project that the effects of these increased concentrations will include the following:

- Warmer average temperatures on Earth

- Increased levels of evaporation and precipitation (this will vary by region, with some areas becoming wetter while others become dryer)

- Increased ocean temperatures, which will cause seawater to expand and also lead to increased glacier and sea ice melting, resulting in rising sea levels

- Shifts in location and makeup of natural plant communities

Figure 4.7 provides an overview of 10 climate change indicators, some of which show increasing (white arrows) and others decreasing (black arrows) trends. All these indicators provide evidence and support for the conclusion that global warming is occurring.

Scientists also study climates of the past, known as paleoclimates. The climate on Earth has changed throughout geologic time. For instance, over 300 million years ago, during the Paleozoic Era, the land that is now North America was centered along the equator, and this resulted in a warmer climate. One way scientists learn about paleoclimates is by studying ice cores, which are like little time capsules. Ice cores provide scientists an opportunity to sample the air temperature and chemistry from another time (the time at which the water froze to ice). By studying past concentrations of greenhouse

**Figure 4.7. Ten Indicators of a Warming World**

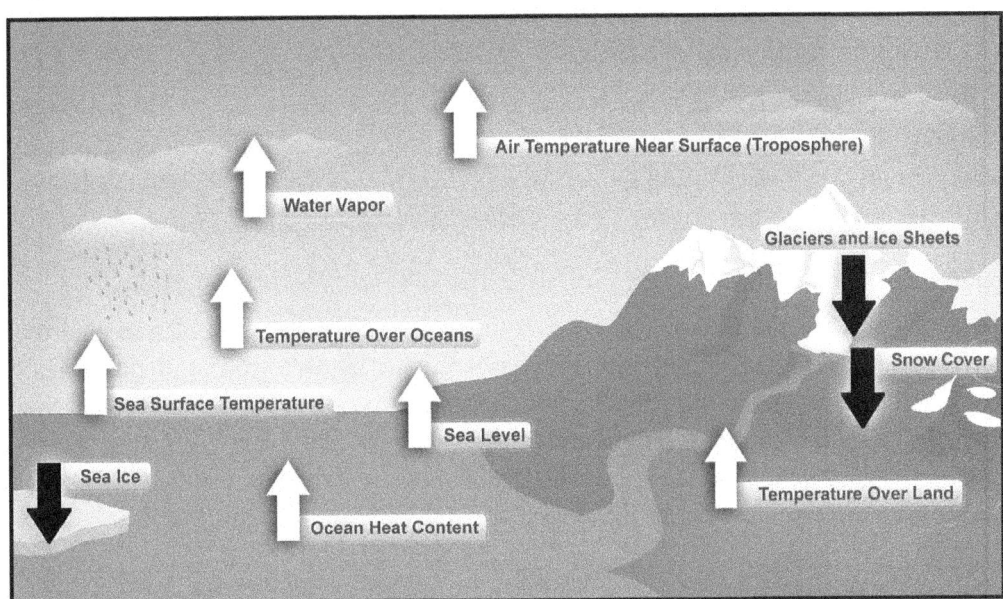

*Source:* U.S. Global Change Research Program, *2014 National Climate Assessment, https://nca2014. globalchange.gov/report/our-changing-climate/observed-change.*

*Note:* A full-color version of this figure is available on the book's Extras page at *www.nsta.org/ roadmap-humanimpacts.*

gases in ice cores, scientists can determine how those concentrations have changed over time. Current technologies allow scientists to collect data from ice cores that are up to 800,000 years old. Other paleoclimate indicators include rocks and fossils.

## The Industrial Revolution and Changing Technologies

Human technological development, beginning with the Industrial Revolution, has caused a large increase in greenhouse gases. To understand human impacts on our climate, it's important to have knowledge about this era, the changes it brought, and their positive and negative effects on society. A number of excellent teacher resources are available to support the exploration of the Industrial Revolution:

- This article provides a number of useful links: *www.theguardian.com/teacher-network/2012/aug/06/industrial-revolution-teaching-resources.*

- Infographics such as this one are useful for supporting discussion: *www. historycrunch.com/industrial-revolution-overview-infographic.html#.*

- Brief, engaging video overviews are useful as well, such as this one from Khan Academy: *www.khanacademy.org/partner-content/big-history-project/acceleration/bhp-acceleration/v/bhp-industrial-revolution-crashcourse*.

Technologies have continued to change up to the present day, with several periods of technology-driven changes in society that can be thought of as multiple industrial revolutions. A brief, useful teacher background piece can be found at *www.cnbc.com/2019/01/16/fourth-industrial-revolution-explained-davos-2019.html*.

## Jigsaw Method

The jigsaw method is a research-based instructional strategy that promotes cooperative learning in the classroom. When using the jigsaw method in this lesson, students research a single climate change indictor together in one small group, and then explain what they learned about this indicator to a different group. In this way, it is like putting together the pieces of a jigsaw puzzle so that all students gain a more complete understanding of the subject. More information on the jigsaw method can be found at *www.jigsaw.org*.

## COMMON MISCONCEPTIONS

Students will have various types of prior knowledge about the concepts introduced in this lesson. Table 4.7 outlines some common misconceptions students may have concerning these concepts. Because of the breadth of students' experiences, it is not possible to anticipate every misconception that students may bring as they approach this lesson. Incorrect or inaccurate prior understanding of concepts can influence student learning in the future, however, so it is important to be alert to misconceptions such as those presented in the table.

**Table 4.7. Common Misconceptions About the Concepts in Lesson 2**

| Topic | Student Misconception | Explanation |
|---|---|---|
| Greenhouse effect | Greenhouse gases are only caused by human activity. | Greenhouse gases are a natural and necessary part of Earth's systems. Without the greenhouse effect, planet Earth would be too cold for life to exist as we know it. |
| Global warming | Recent global warming is caused by the Sun. | Scientists monitor the Sun with satellites and have found no increase in the amount of solar energy reaching Earth in the past 30 years. |
| | Water vapor in the atmosphere is the greenhouse gas primarily responsible for global warming. | There is an increase in water vapor in the atmosphere, but this is in response to the increased $CO_2$. |

## PREPARATION FOR LESSON 2

Review the Teacher Background Information section (p. 80) and preview the recommended videos in the Learning Components section to ensure that you have the necessary foundational knowledge about the greenhouse effect, greenhouse gases, and climate change indicators to teach this lesson. Assemble the materials for the lesson and make copies of student handouts. (A full-color version of the Greenhouse Effect Simulation Student Handout is available on the book's Extras page at *www.nsta.org/roadmap-humanimpacts.*) Download the greenhouse effect simulation at *https://phet.colorado.edu/en/simulation/legacy/greenhouse* onto the class computers.

You need to decide beforehand which climate change indicators you feel would be best for students to research in the jigsaw activity. Consider your community: Which indicators may have more meaning for your students? Visit the EPA's climate change indicators report at *www.epa.gov/climate-indicators/printer-friendly-pdf-downloads-indicator-text-and-figures* for printer-friendly versions of these indicators. Each student in a group should have a copy of the assigned climate change indicator. Depending on the technology you have available, you can provide printed handouts or have students access these electronically. Full-color versions of all NOAA climate change indicator graphs are available on the book's Extras page at *www.nsta.org/roadmap-humanimpacts.*

## LEARNING COMPONENTS
### Introductory Activity/Engagement

**Connection to the Challenge:** Begin the lesson by directing students' attention to the driving question for the module and challenge: How can we develop a local response to address an aspect of human impact on global climate change? Hold a brief discussion of how what they learned in Lesson 1 has influenced their thinking regarding climate change. Review the list of students' climate change questions created in Lesson 1 and have students offer answers to questions based on what they have learned. Note students' responses on the pages you created for questions.

**Science Class:** Ask students, "On a night when it is cold, how do you stay warm when you go to sleep?" Gather students' responses. Then ask students, "Why do we wear coats on cold days?" Gather responses. Have students discuss in pairs, "How do your bed covers or your jacket help keep you warm?" After students have had a moment to discuss, have them share their thinking. Sample responses might be that their coats and bed covers help trap heat or serve as insulators.

**Mathematics Connection:** Not applicable.

**ELA and Social Studies Connections:** Show students a video of the rise and fall of tides, such as "Fall and Rise of the Tide in the Bay of Fundy at Hall's Harbour, Nova Scotia" at *www.youtube.com/watch?v=OP0cpXpw8yk*.

Ask students what causes the water to rise and fall. If students don't mention tides, share this word with them. Then, show a video of what causes tides, such as "Why Does the Tide Come In and Go Out Again?" at *www.youtube.com/watch?v=m8UGm-dKAoE*.

Show students a map of the United States. Ask students to name some of the states that border an ocean. Then, have students locate Miami, Florida, on the map. Ask if any students have ever visited Miami Beach. Have them share what it is like.

Provide students with a news clip of high tides in Miami Beach, such as "High Tides Cause Flooding Across Miami Beach" at *www.youtube.com/watch?v=GqSr2RF13Pg* or "Miami Beach Residents Say They're Tired of Continued High Tide Flooding" at *www.youtube.com/watch?v=omXnXbTxZjU*.

Show students the video "Sea Level Rise Is So Much More Than Melted Ice" at *www.youtube.com/watch?v=SA5zh3yG_-0*. Then, hold a class discussion on what causes the sea level to rise. Ask students which of these causes are human impacts and which are natural processes.

Have students do internet research to explore some ways that coastal cities are addressing sea level rise. Then, hold a class discussion about how sea level rise will affect society, asking questions such as the following:

- What are some of the consequences of sea level rise on coastal cities around the world?

- What other places in the world are threatened by sea level rise?

- What are some ways that people are engineering to minimize the impact on coastal cities?

## Activity/Exploration

**Science Class:** Tell students: "Earlier, we learned about the layers of Earth's atmosphere and how the majority of weather occurs in the lowermost layers. Today we are going to begin studying the gases that make up our atmosphere and the role they play in Earth's weather and climate." Tell them that they will explore the greenhouse effect by working in teams of three or four to build models of greenhouses. Provide students with copies of the Greenhouse Model Setup Procedures and Activity Sheet handouts to put in their STEM Research Notebooks. Have student teams follow the setup procedures to make their models. Then, students should individually record the temperatures of their team's bottles on the Greenhouse Model Activity Sheet every 3 minutes for 24 minutes. Next, students should make a graph of their data and analyze the trend for each bottle. Have

students write a sentence summarizing their findings. Students should notice that the bottle with the sodium bicarbonate reached a higher temperature over time. *Note:* Do not tell students what gas is collecting in the bottle at this time, as they will learn more about greenhouse gases, in particular carbon dioxide.

The next day, have students begin to explore the greenhouse effect simulation at *https://phet.colorado.edu/en/simulation/greenhouse,* using the Greenhouse Effect Simulation Student Handout to guide them. Students may work individually or in small groups of two or three. Break this activity up over several days (see Table 3.7 on p. 38 for suggestions on how much to complete each day). Have students complete Parts 1 and 2 of the handout the first day they work with the simulation.

At the end of class, after students have completed Part 2, have them respond to the following on index cards and turn them in as exit tickets: *Which gases from the simulation are greenhouse gases? How do you know?* Use their answers to check for student understanding, and correct any misconceptions before moving on.

During the next two science classes, have students return to the simulation and work on Parts 3–5. From the simulation, students learn how a greenhouse traps heat inside its glass walls, whether all greenhouse gases contribute to the greenhouse effect, and which gases from the simulation are greenhouse gases. They also examine the amount of $CO_2$ in the atmosphere and the average temperature at different points in Earth's history. Finally, students make conclusions about the role of greenhouse gases in Earth's atmosphere, predict what will happen to average global temperature given the trend of increased $CO_2$ in the atmosphere, and synthesize data from graphing the global temperature averages from Lesson 1 with the information they learned about greenhouse gases in this lesson.

**Mathematics Connection:** Review how to construct a scatter plot with students. Emphasizing the need to clearly label the axes and include units. Consider giving them a small set of data points to construct a scatter plot. Then either discuss with them how to extrapolate from their data or use a video such as "Making Predictions on a Scatter Plot Using Interpolation and Extrapolation" at *www.youtube.com/watch?v=bEANDlJkqcU* to help students understand how to make predictions for the temperature of their greenhouse models.

**ELA and Social Studies Connections:** Have students explore the potential effects of global warming on communities across the United States. Group students into teams of three or four students and, using a U.S. map, have each team choose a community from a different region of the nation. Have teams conduct internet research to determine how climate change could impact this community over the next 100 years. Internet resources such as the following may be helpful to students:

- "America After Climate Change, Mapped" at *www.citylab.com/environment/2019/ 12/green-new-deal-atlas-climate-change-mcharg-center-maps-model/603415*

- The Climate Central webpage at *https://www.climatecentral.org/gallery/maps*

- The NOAA Climate webpage at *htttps://www.climate.gov*

Students should propose possible solutions for helping these communities deal with the effects of climate change and think about what they can do locally to prevent global climate change. Ask students to create multimedia presentations such as stop-motion animation to present their findings. You should ensure that students are using a variety of credible sources and that their proposed solution may be able to influence the global human population.

Also, use current events to help students make a connection to the politics of climate change. For instance, students could research Greta Thunberg to learn how someone their own age is making a big impact on climate science awareness. Facts about Greta Thunberg can be found at *www.natgeokids.com/nz/kids-club/cool-kids/general-kids-club/ greta-thunberg-facts*.

## Explanation

**Science Class:** Review students' work from the greenhouse effect simulation. Turn the conversation to the greenhouse models students created earlier. Tell them that the sodium bicarbonate tablets released $CO_2$ gas into the bottle. Ask students how the data from their greenhouse models correlate with what they learned from the simulation.

Have students watch a video on the greenhouse effect, such as the animation at *https:// climate.nasa.gov/causes*. Then, have them watch a video about the sources of $CO_2$ in the atmosphere, such as "Climate 101: Causes and Effects" at *www.nationalgeographic.com/ environment/global-warming/greenhouse-gases* or "Climate Change: Evidence and Causes" at *https://royalsociety.org/topics-policy/projects/climate-change-evidence-causes*.

As students watch the videos, ask them to take notes in their STEM Research Notebooks about the sources of $CO_2$ in the atmosphere, the causes for the increase in $CO_2$, and the effects caused by the increase in $CO_2$. Hold a class discussion about students' main takeaway from the videos. Ultimately, students need to recognize that since the Industrial Revolution, humans have been producing extreme amounts of $CO_2$ and releasing it into the atmosphere. To conclude the discussion, have students watch the video "$CO_2$ and Temperature" at *https://climate.nasa.gov/ask-climate*.

**Mathematics Connection:** Not applicable.

**ELA Connection:** Expand discussion of the climate data graphic in Lesson 1 (see p. 51) with a study of the Industrial Revolution, examining how it changed society in positive and negative ways. Discuss how people corrected problems that resulted, including social, political, and technological inventions. Then, discuss how industry is continually revolutionizing, examining the idea of multiple industrial revolutions rather than just the one we commonly identify—the first fueled by invention, the next mass production, the third digitization, and the current one technology (see p. 83 in the Teacher Background Information section). Consider together how innovation today has the potential to both solve and create problems for society.

**Social Studies Connection:** Have students give their multimedia presentations. Consider formatting the presentations as a community event with guests from the community.

## Elaboration/Application of Knowledge

**Science Class:** Using the jigsaw method (see p. 84 of Teacher Background Information), divide students into small groups of four or five and assign each group a single climate change indicator out of the ones you decided to use from the EPA's Climate Change Indicator website during the lesson preparation. To scaffold students' learning, choose an indicator from the list below to review with the whole class. There is a lot of information in these reports, so as you go over a sample report, point out the things students should pay close attention to.

The website has many indicators to choose from, but the following may be easiest for students to grasp:

- Arctic Sea Ice

- Atmospheric Concentrations of Greenhouse Gases

- Bird Wintering Ranges

- Cherry Blossom Bloom Dates in Washington, DC

- High and Low Temperatures

- Leaf and Bloom Dates

- Ocean Heat

- Sea Surface Temperature

- U.S. and Global Temperatures

- U.S. Greenhouse Gas Emissions

Remind the class that every student in a group is responsible for being able to summarize the group's assigned indicator, as they will be the experts on that indicator in the next group they are assigned to. Each group should summarize the following information from the report:

- What is the indicator?

- According to the report, what is the trend of the indicator?

- What are the key points made in the report?

Once the indicator expert groups have had time to study their assigned indicators, renumber each student in a group; this is their new group number. Students now go to their new groups, where each one serves as an expert to report on a different indicator to their new group. Students should record the findings on each indicator in their STEM Research Notebooks.

### STEM Research Notebook Prompt

Have students respond to the following questions in their STEM Research Notebooks:

- *What is a climate change indicator?*

- *What are general trends that you have noticed in the indicators you studied?*

- *What do the data trends indicate to you about the state of Earth's climate?*

- *What evidence do you have to support that Earth's climate is changing?*

**Mathematics and ELA Connections:** Divide students into groups of three or four, and assign each group one of the NOAA Climate Change Indicator Graphs (p. 107) to develop a claim, evidence, reasoning (CER) response from the trend of the data and discuss the influence on our climate. Have students put copies of the graphs in their STEM Research Notebooks and record their group's claim on a Claim, Evidence, Reasoning student handout (see Lesson 1, p. 66). Hold a whole-class discussion to allow students to share. Have students record other groups' inferences on the appropriate graphs.

**Social Studies Connection:** Have students investigate how climate change affects different sectors of society (e.g., human health, society, transportation, economics). Encourage them to think about local, regional, national, and international levels.

## Evaluation/Assessment

Students may be assessed on the following performance tasks and other measures listed.

*Performance Tasks*

- Greenhouse Model Activity Sheet (p. 96)

- Greenhouse Effect Simulation Student Handout (p. 98)

- Multimedia presentation

- Claim, Evidence, Reasoning Student Handout (see Lesson 1, p. 66)

*Other Measures*

- Collaboration Rubric (see Lesson 1, p. 73)

- STEM Research Notebook Entry Rubric (see Lesson 1, p. 72)

- Participation in class discussions

## INTERNET RESOURCES

"Changing Planet: Past, Present, Future" lecture series
- *www.hhmi.org/biointeractive/changing-planet-past-present-future*

"Climate Change: How Do We Know We're Not Wrong?" lecture
- *www.hhmi.org/biointeractive/climate-change-how-do-we-know-were-not-wrong*

Climate change indicators
- *www.epa.gov/climate-indicators*

"Global Climate Change: Vital Signs of the Planet"
- *http://climate.nasa.gov*

Satellite images of climate change
- *https://climate.nasa.gov/images-of-change*

Industrial Revolution
- *www.theguardian.com/teacher-network/2012/aug/06/industrial-revolution-teaching-resources*

- *www.historycrunch.com/industrial-revolution-overview-infographic.html#*

- *www.khanacademy.org/partner-content/big-history-project/acceleration/bhp-acceleration/v/bhp-industrial-revolution-crashcourse*

Changing technologies
- *www.cnbc.com/2019/01/16/fourth-industrial-revolution-explained-davos-2019.html*

*A Framework for K–12 Science Education: Practices, Crosscutting Concepts, and Core Ideas*
- *www.nap.edu/catalog/13165/a-framework-for-k-12-science-education-practices-crosscutting-concepts*

*Taking Science to School: Learning and Teaching Science in Grades K–8*
- *www.nap.edu/catalog/11625/taking-science-to-school-learning-and-teaching-science-in-grades*

Jigsaw method
- *www.jigsaw.org*

Climate change indicators
- *www.epa.gov/climate-indicators/printer-friendly-pdf-downloads-indicator-text-and-figures*

"Fall and Rise of the Tide in the Bay of Fundy at Hall's Harbour, Nova Scotia" video
- *www.youtube.com/watch?v=OP0cpXpw8yk*

"Why Does the Tide Come In and Go Out Again?" video
- *www.youtube.com/watch?v=m8UGm-dKAoE*

"High Tides Cause Flooding Across Miami Beach" video
- *www.youtube.com/watch?v=GqSr2RF13Pg*

"Miami Beach Residents Say They're Tired of Continued High Tide Flooding" video
- *www.youtube.com/watch?v=omXnXbTxZjU*

"Sea Level Rise Is So Much More Than Melting Ice" video
- *www.youtube.com/watch?v=SA5zh3yG_-0*

"Making Predictions on a Scatter Plot Using Interpolation and Extrapolation" video
- *www.youtube.com/watch?v=bEANDlJkqcU*

Resources for the effects of global warming on communities:
- "America After Climate Change, Mapped," *www.citylab.com/environment/2019/12/green-new-deal-atlas-climate-change-mcharg-center-maps-model/603415*
- Climate Central webpage, *www.climatecentral.org/gallery/maps*
- NOAA Climate webpage, *www.climate.gov*

Greta Thunberg facts
- *www.natgeokids.com/nz/kids-club/cool-kids/general-kids-club/greta-thunberg-facts*

Greenhouse effect animation
- *https://climate.nasa.gov/causes*

"Climate 101: Causes and Effects" video
- *www.nationalgeographic.com.au/videos/other/climate-101-causes-and-effects-5977.aspx*

"Climate Change: Evidence and Causes" video
- *https://royalsociety.org/topics-policy/projects/climate-change-evidence-causes*

"CO$_2$ and Temperature" video
- *https://climate.nasa.gov/ask-climate*

Name: _____    Date: _____

**STUDENT HANDOUT, PAGE 1**

# GREENHOUSE MODEL SETUP PROCEDURES

In this activity, you will work with your team to build a model of a greenhouse to help you learn about the greenhouse effect. A greenhouse is usually built out of glass walls. The glass walls allow for light, heat, and energy from the Sun to pass through, which keeps the air and the plants warm. None of this warm air can escape, because it is trapped within the glass walls. Your models use clear plastic bottles to simulate greenhouses.

## Procedure

1. Pour 150 milliliters (mL) of water into each bottle.

2. Label one bottle "Control" and the other bottle "Gas."

3. Make a plug for each bottle from a lump of clay. Gently insert a thermometer into each bottle through the clay plug. The end of the thermometer should not be in the water but should be hanging in the air above the water.

4. Position the bottles in front of the heat lamp, making sure that each bottle is the same distance away.

5. Remove the plug from the bottle labeled "Gas." Place the sodium bicarbonate tablets into the bottle and replace the plug.

6. Immediately cover the top of each bottle (including the thermometer) with a piece of plastic wrap. Secure the plastic wrap with a rubber band.

7. Take and record an initial temperature reading.

8. Turn on the heat lamp.

9. Record the temperature of each bottle every 3 minutes on your Greenhouse Model Activity Sheet.

Name: _____     Date: _____

# GREENHOUSE MODEL SETUP PROCEDURES

## Safety Notes

1. All students must wear safety goggles and nonlatex aprons during the setup, hands-on, and takedown segments of the activity.

2. Use caution when working with glass or plasticware, which can break and cut skin.

3. Immediately pick up any items dropped on the floor to avoid a slip-and-fall hazard.

4. Immediately wipe up any spilled water on the floor to avoid a slip-and-fall hazard.

5. Use caution when working with the heat lamp, which can burn skin on medium and high settings.

6. Use only GFI-protected circuits when using electrical equipment, and keep electrical wires away from water sources to avoid risk of shock.

7. Do not place the tablets used in the lab activity in your mouth or swallow them.

8. Wash hands with soap and water after completing this activity.

Name: _____          Date: _____

**STUDENT HANDOUT, PAGE 1**

## GREENHOUSE MODEL ACTIVITY SHEET

Use your greenhouse model to complete this activity sheet.

1. Record the temperature every 3 minutes over a period of 24 minutes for each of your bottles. Be sure to indicate whether you are measuring in degrees Celsius or Fahrenheit.

| Time (minutes) | | | | | | | | |
|---|---|---|---|---|---|---|---|---|
| Control (temperature) | | | | | | | | |
| Gas (temperature) | | | | | | | | |

Name: _____     Date: _____

## GREENHOUSE MODEL ACTIVITY SHEET

2. Graph your data below. Use a different color to record the temperature for each bottle across time. Include a legend and clearly label each axis.

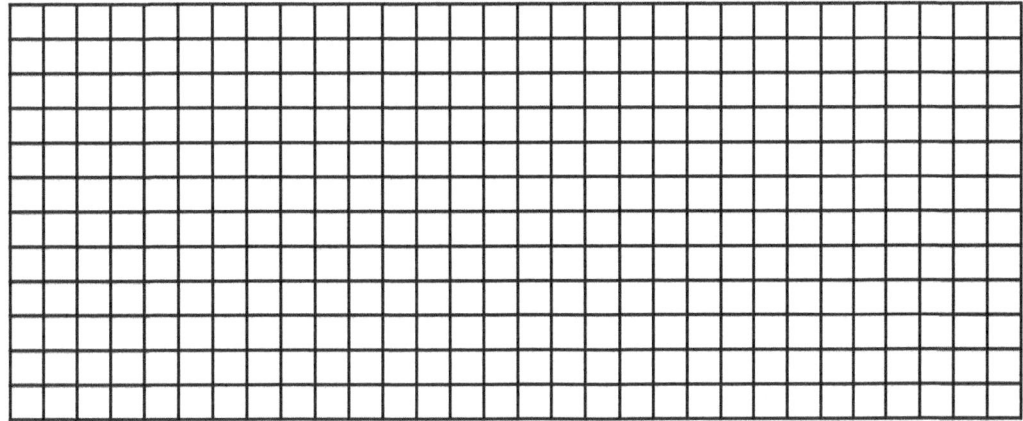

3. Write a sentence describing what you notice.

Name: _____  Date: _____

**STUDENT HANDOUT, PAGE 1**

## GREENHOUSE EFFECT SIMULATION

In this activity, you will use an interactive simulation to learn about learn about greenhouse gases and their behaviors in the atmosphere, as well as how the greenhouse effect works. Use this handout along with the simulation found at *https://phet.colorado.edu/en/simulation/legacy/greenhouse*.

*Note:* Full-color versions of the images in this handout are available on the book's Extras page at *www.nsta.org/roadmap-humanimpacts*.

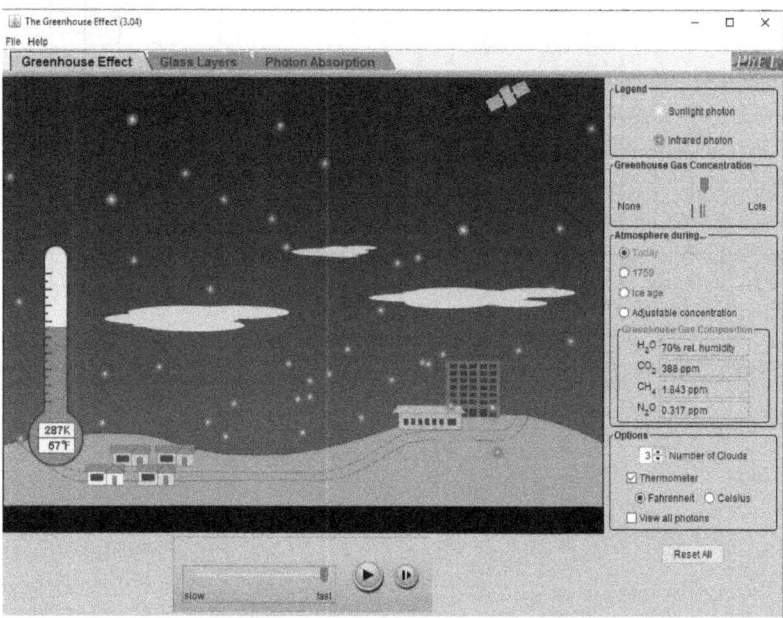

### Part 1. Become Familiar With the Simulation

1. Look over the different features of the simulation.

    a. Notice the legend.
    b. Notice the sunlight photons and infrared photons that are moving on the screen. Photons are particles of electromagnetic energy or light. The sunlight photons (yellow) represent energy that is radiating from the Sun. The infrared photons (red) represent the energy (heat) that is reflected from Earth and reenters the atmosphere.
    c. Notice the slider bar at the bottom of the simulation, which lets you slow down or speed up the movement of the photons.

**4**

Name: _____

Date: _____

**STUDENT HANDOUT, PAGE 2**

# GREENHOUSE EFFECT SIMULATION

2. Under Options, you are able to change the number of clouds in the sky. What happens when you increase the number of clouds in the sky? Jot down what you notice.

3. On the right side of the simulation is the phrase "Atmosphere during ..." with several options below it to see simulations of the atmosphere today, in the year 1750, and during the Ice Age. Use this feature to answer the following questions:

   a. What is the average temperature today?

   b. What was the average temperature during the Ice Age?

   c. Do you notice more infrared photons or sunlight photons when you look at the simulation of today?

Name: _____     Date: _____

## GREENHOUSE EFFECT SIMULATION

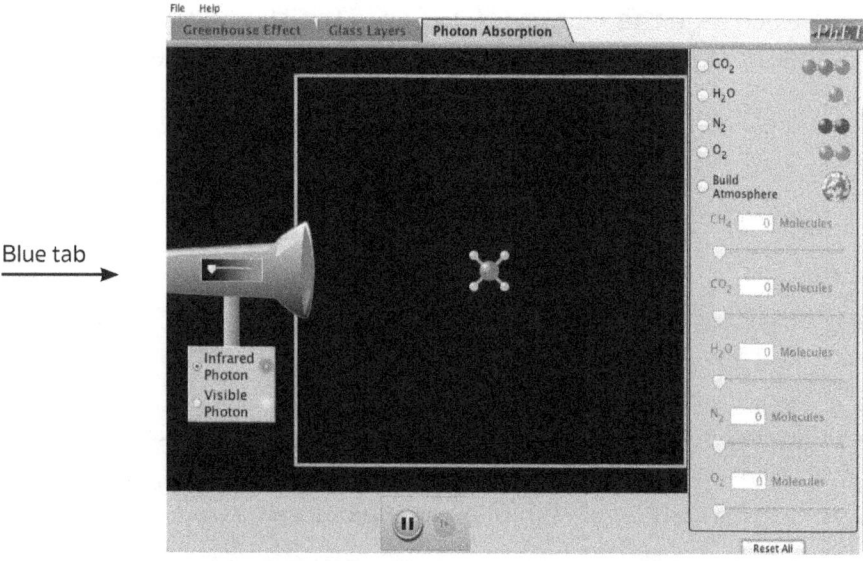

Blue tab →

### Part 2. Determine Which Gases Are Greenhouse Gases

1. Click on the Photon Absorption tab. This simulation represents a glass house. Record your findings in the table on the next page.

2. Fully increase the speed of photon release by sliding the blue tab all the way to the right.

3. Select Infrared Photon and watch the photon pass through each of the gases listed on the right-hand side for 10 seconds. If an Infrared Photon is absorbed by the gas (you will see the gas molecule move), then write Yes in the first column of the table below. If a gas allows the photon to freely pass, then write No. Repeat for the Visible Photon, writing Yes or No in the second column.

4. Let's assume that the glass house is the planet Earth. Using the definition of greenhouse gases (gases that absorb infrared radiation, trap heat [energy] from the Sun in the atmosphere, and contribute to the greenhouse effect), which of the gases are greenhouse gases? Write Yes or No for each gas in the third column of the table.

4

Name: _____     Date: _____

**STUDENT HANDOUT, PAGE 4**

## GREENHOUSE EFFECT SIMULATION

| Atmospheric Gas | Absorbed Infrared Photon (Yes or No) | Absorbed Visible Photon (Yes or No) | Greenhouse Gas (Yes or No) |
|---|---|---|---|
| Methand ($CH_4$) | | | |
| Carbon dioxide ($CO_2$) | | | |
| Water ($H_2O$) | | | |
| Nitrogen ($N_2$) | | | |
| Oxygen ($O_2$) | | | |

Name: _____     Date: _____

# GREENHOUSE EFFECT SIMULATION

**Part 3. Explore Atmospheric Temperature and Numbers of Glass Panes Photons Travel Through**

1. Click on the Glass Layers tab. This simulation allows you to study atmospheric temperature and different numbers of glass panes in a glass house for the photons to travel through.

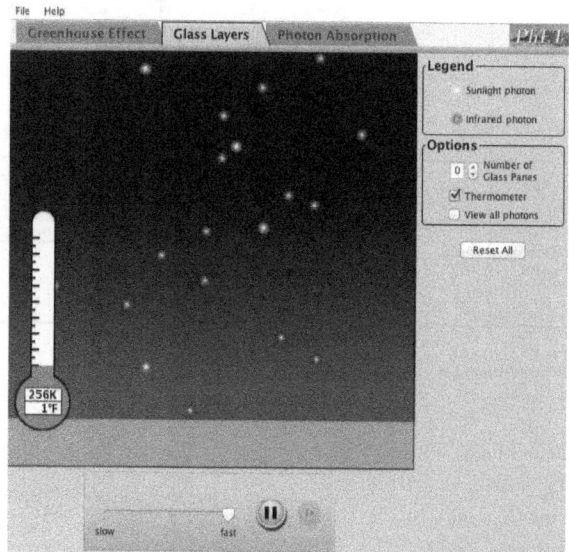

2. Alter the number of glass panes under Options and record your data in the table below. Then, use your data to plot a graph. (*Hint:* The number of glass panes is your independent variable.) Use your data to determine a scale for both the *x*-axis and the *y*-axis.

Name: _____     Date: _____

**STUDENT HANDOUT, PAGE 6**

## GREENHOUSE EFFECT SIMULATION

| Number of Glass Panes | Temperature (°F) |
| --- | --- |
| | |
| | |
| | |
| | |

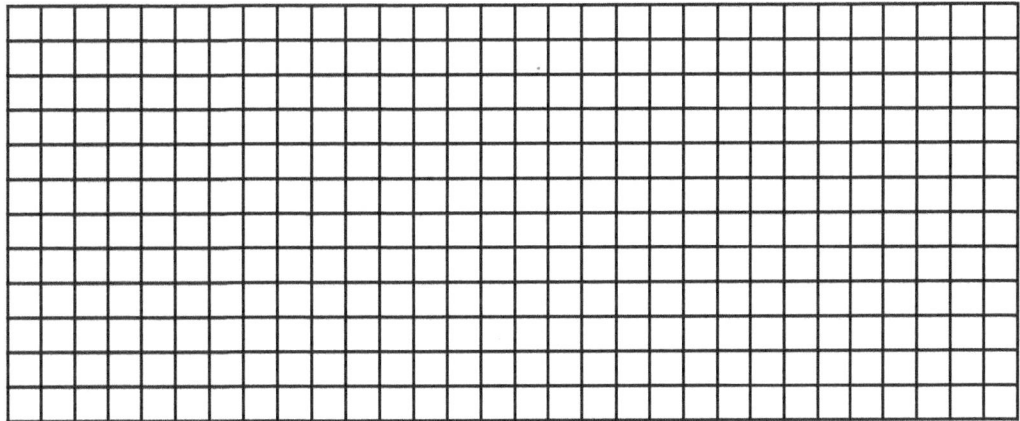

Human Impacts on Our Climate Lesson Plans

Name: _____     Date: _____

**STUDENT HANDOUT, PAGE 7**

# GREENHOUSE EFFECT SIMULATION

3. Based on your data, what trend do you notice?

4. Describe what happens to the following as you add more glass panes:

   a. The movement of the sunlight photons to Earth

   b. The absorption of the infrared photons

Name: _____     Date: _____

**STUDENT HANDOUT, PAGE 8**

## GREENHOUSE EFFECT SIMULATION

**Part 4. Explore the Greenhouse Effect in Different Eras and Investigate the Effects of Clouds**

1. Return to the Greenhouse Effect tab. Click on each of the time periods listed under "Atmosphere during …" to determine the amount of $CO_2$ in the atmosphere and the average temperature during that time. (*Note:* For "none," slide the concentration bar all the way to the left.)

| Greenhouse Gas Concentration | Amount of Greenhouse Gas $CO_2$ (ppm) | Average Temperature (°F) |
|---|---|---|
| None | | |
| Ice Age | | |
| 1750 | | |
| Today | | |

2. Based on your data, what do you predict the average temperature will be in 100 years? The amount of $CO_2$?




3. What would happen if greenhouse gases did not exist in the atmosphere?




4. After clicking the Reset All button, study the motion of the sunlight and infrared protons as you add a cloud every 20 seconds.
   a. Describe the motion of the sunlight photons with the clouds.


   b. Describe the motion of the infrared photons with the clouds.


5. Clouds are made from water vapor (gas form of water) that has risen in the air, gets cooled, and becomes small droplets of water or ice that are able to stay in the air because of their small size. Based on your data, are clouds keeping Earth cooler or warmer?

Name: _____     Date: _____

# GREENHOUSE EFFECT SIMULATION

**Part 5. Consider the Effects of Greenhouse Gases Based on What You Learned From the Simulation**

1. How do greenhouse gases affect our climate?

2. Do we need greenhouse gases in our atmosphere?

3. Look at the image below. This figure provides 800,000 years of carbon dioxide levels. Given this data and what you learned about greenhouse gases, what would you predict will happen to Earth's average global temperature over the time period covered in the graph?

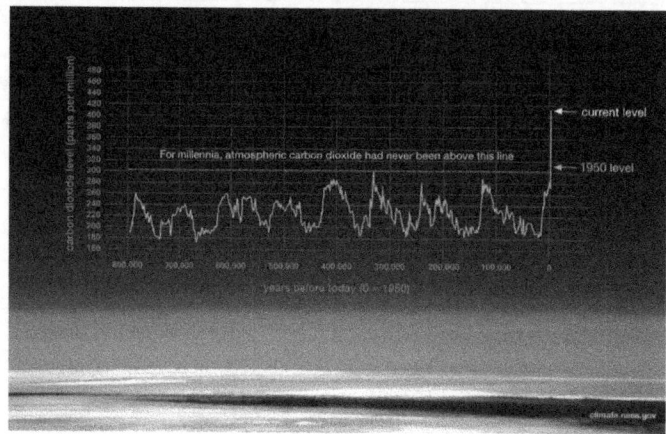

*Source: https://climate.nasa.gov/climate_resources/24/graphic-the-relentless-rise-of-carbon-dioxide.*

4. Based on your graphing activity on average global temperatures in Lesson 1 and what you have learned from the simulation about greenhouse gases, what conclusions can you make about what is causing global climate change?

Name: _____     Date: _____

**STUDENT HANDOUT, PAGE 1**

# NOAA CLIMATE CHANGE INDICATOR GRAPHS

*Note:* Full-color versions of the images in this handout are available on the book's Extras page at *www.nsta.org/roadmap-humanimpacts*.

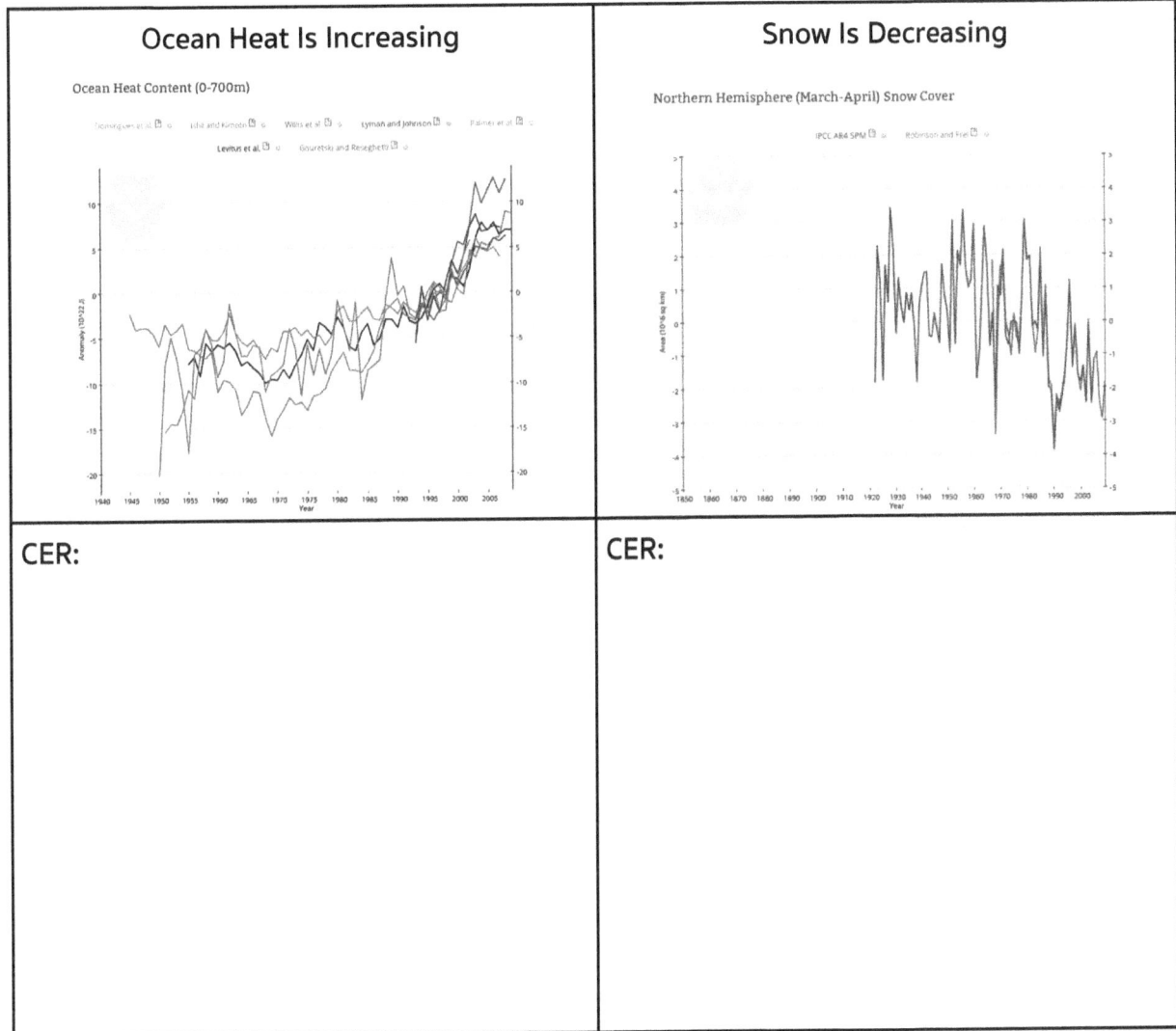

**Ocean Heat Is Increasing**

Ocean Heat Content (0-700m)

**Snow Is Decreasing**

Northern Hemisphere (March-April) Snow Cover

CER:

CER:

*Source:* Graphs from *https://cpo.noaa.gov/warmingworld/ocean_heat_content.html, https://cpo.noaa.gov/warmingworld/snow.html.*

*Note:* CER = claim, evidence, reasoning.

Name: _____     Date: _____

**STUDENT HANDOUT, PAGE 2**

# NOAA CLIMATE CHANGE INDICATOR GRAPHS

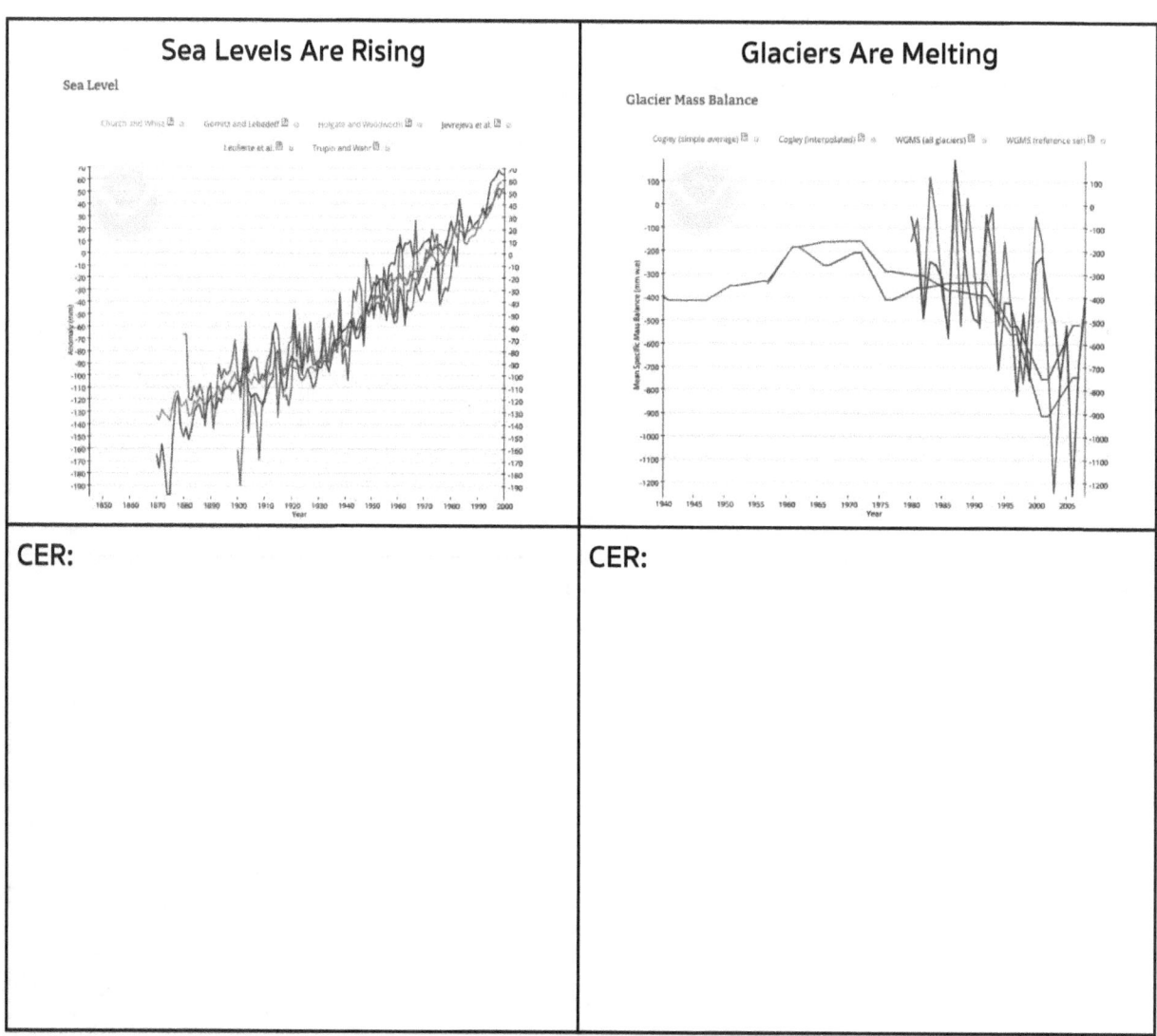

| Sea Levels Are Rising | Glaciers Are Melting |
|---|---|
| Sea Level | Glacier Mass Balance |
| **CER:** | **CER:** |

*Source:* Graphs from *https://cpo.noaa.gov/warmingworld/ocean_heat content.html, https://cpo.noaa.gov/warmingworld/snow.html., https://cpo.noaa.gov/warmingworld/global_sea_level.html,* and *https://cpo.noaa.gov/warmingworld/glaciers.html.*

*Note:* CER = claim, evidence, reasoning.

Name: _____     Date: _____

# NOAA CLIMATE CHANGE INDICATOR GRAPHS

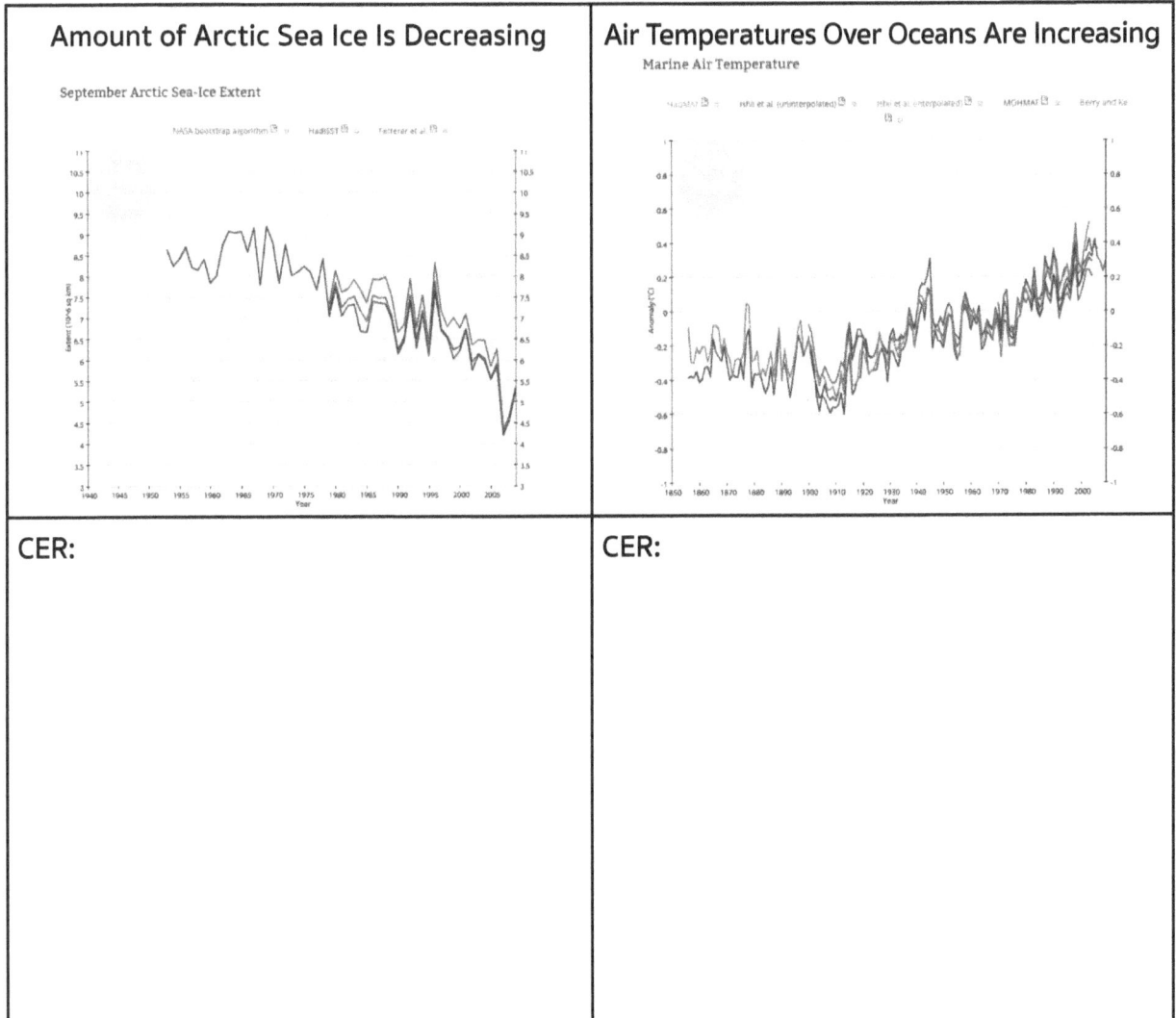

### Amount of Arctic Sea Ice Is Decreasing

September Arctic Sea-Ice Extent

CER:

### Air Temperatures Over Oceans Are Increasing

Marine Air Temperature

CER:

*Source:* Graphs from *https://cpo.noaa.gov/warmingworld/arctic_sea_ice.html, https://cpo.noaa.gov/warmingworld/air_temperature_over_ocean.html.*

*Note:* CER = claim, evidence, reasoning.

Name: _____     Date: _____

**STUDENT HANDOUT, PAGE 4**

# NOAA CLIMATE CHANGE INDICATOR GRAPHS

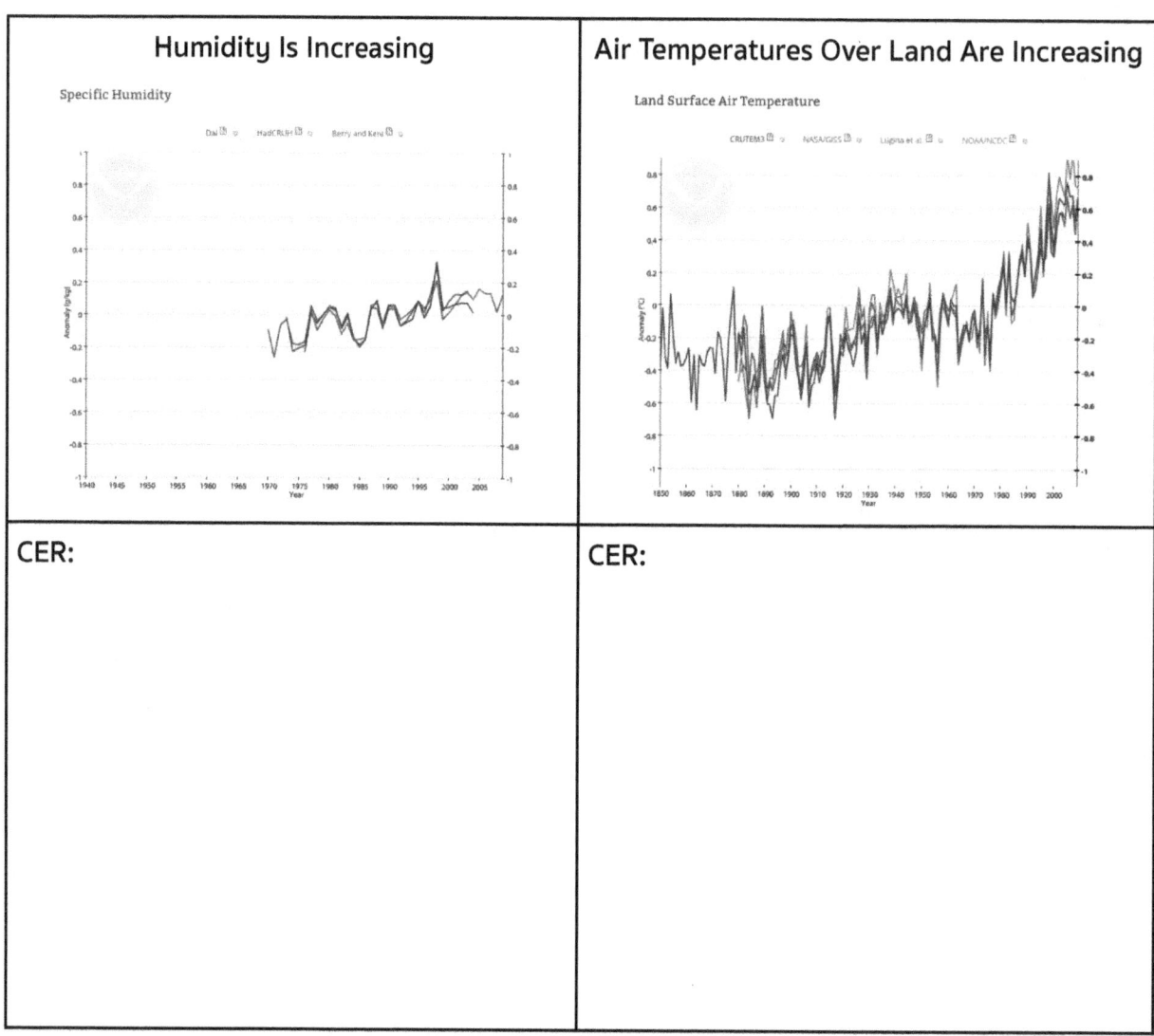

| Humidity Is Increasing | Air Temperatures Over Land Are Increasing |
|---|---|
| **CER:** | **CER:** |

*Source:* Graphs from *https://cpo.noaa.gov/warmingworld/humidity.html,* and *https://cpo.noaa.gov/warmingworld/ air_temperature_over_land.html.*

*Note:* CER = claim, evidence, reasoning.

Name: _____     Date: _____

**STUDENT HANDOUT, PAGE 5**

# NOAA CLIMATE CHANGE INDICATOR GRAPHS

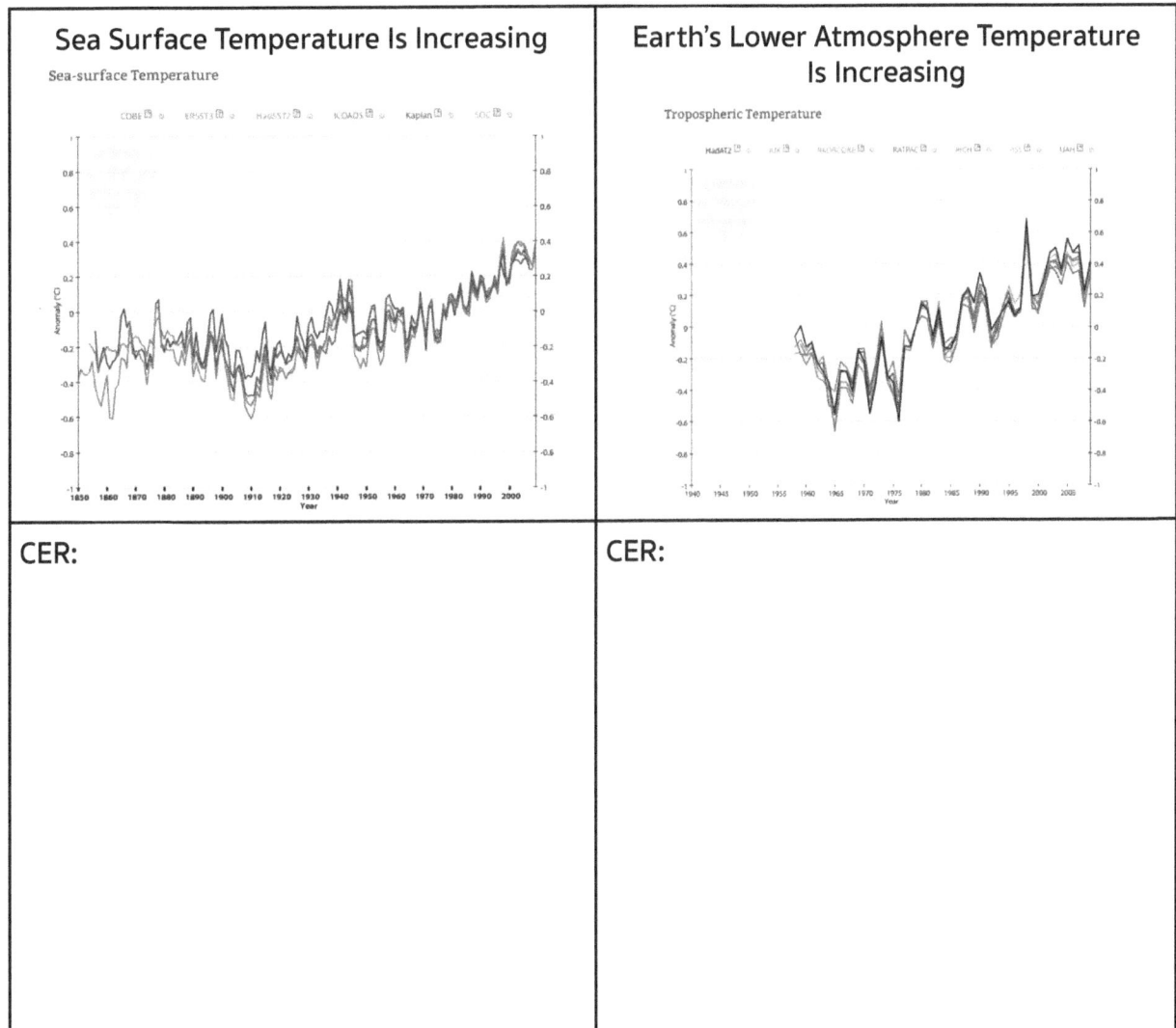

### Sea Surface Temperature Is Increasing

Sea-surface Temperature

### Earth's Lower Atmosphere Temperature Is Increasing

Tropospheric Temperature

**CER:**

**CER:**

*Source:* Graphs from *https://cpo.noaa.gov/warmingworld/sea_surface_temperature.html* and *https://cpo.noaa.gov/warmingworld/temperature_of_the_lower_atmosphere.html*.

*Note:* CER = claim, evidence, reasoning.

# Lesson Plan 3: Reducing Your Carbon Footprint

Students learn about the effect climate change is having on pollinators and, as a result, on the natural cycles of plants. Then, they examine trends in other climate change indicators. Students calculate their carbon footprints and design solutions to reduce carbon emissions at the personal, school, or community level.

## ESSENTIAL QUESTIONS

- How does climate affect organisms?
- What is a carbon footprint?
- How can you minimize your carbon footprint?

## ESTABLISHED GOALS AND OBJECTIVES

At the conclusion of this lesson, students will be able to do the following:

- Calculate their own and their households' carbon footprints
- Describe the importance of reducing their carbon footprints
- Understand how climate change affects other organisms and why this is important to humans
- Describe how climate change is influencing the cyclical patterns of different organisms
- Synthesize their understanding of climate change to identify a climate concern in their own lives or in their school or community
- Design a solution to reduce the carbon footprint associated with the identified climate concern

## TIME REQUIRED

- 15 days (approximately 45 minutes each day; see Tables 3.8–3.10, pp. 39–40)

## MATERIALS

*Required Materials for Lesson 3*

- STEM Research Notebooks
- Computer and projector for students to watch videos

- Computers, tablets, or laptops with internet access for student research and presentations (at least 1 per pair of students)

- Handouts (attached at the end of this lesson)

- Writing utensils of various colors (at least two different colors)

- Books

  - *Bees: A Honeyed History,* by Piotr Socha (Harry N. Abrams, 2017)

  - *Energy Island: How One Community Harnessed the Wind and Changed Their World,* by Allan Drummond (Farrar, Straus and Giroux, 2011)

  - *The Bee Book,* by Charlotte Milner (DK Children, 2018)

  - *The Hive Detectives: Chronicle of a Honey Bee Catastrophe,* by Loree Griffin Burns (Scholastic, 2014)

  - *The Honeybee,* by Kirsten Hall (Atheneum Books for Young Readers, 2018)

## CONTENT STANDARDS AND KEY VOCABULARY

Table 4.8 lists the content standards from the *NGSS, CCSS,* and the Framework for 21st Century Learning that this lesson addresses, and Table 4.9 (p. 117) presents the key vocabulary. Vocabulary terms are provided for both teacher and student use. Teachers may choose to introduce some or all of the terms to students.

**Table 4.8. Content Standards Addressed in STEM Road Map Module Lesson 3**

| *NEXT GENERATION SCIENCE STANDARDS* |
| --- |
| **PERFORMANCE EXPECTATIONS** |
| • MS-ESS3-3. Apply scientific principles to design a method for monitoring and minimizing a human impact on the environment. |
| • MS-ESS3-4. Construct an argument supported by evidence for how increases in human population and per-capita consumption of natural resources impact Earth's systems. |
| • MS-ESS3-5. Ask questions to clarify evidence of the factors that have caused the rise in global temperatures over the past century. |
| • MS-ETS1. Define the criteria and constraints of a design problem with sufficient precision to ensure a successful solution, taking into account relevant scientific principles and potential impacts on people and the natural environment. |

*Continued*

**Table 4.8.** (*continued*)

---

### SCIENCE AND ENGINEERING PRACTICES

*Asking Questions and Defining Problems*

Asking questions and defining problems in grades 6–8 builds on grades K–5 experiences and progresses to specifying relationships between variables, and clarifying arguments and models.
- Ask questions to identify and clarify evidence of an argument.

*Analyzing and Interpreting Data*

Analyzing data in 6–8 builds on K–5 and progresses to extending quantitative analysis to investigations, distinguishing between correlation and causation, and basic statistical techniques of data and error analysis.
- Analyze and interpret data to determine similarities and differences in findings.

*Constructing Explanations and Designing Solutions*

Constructing explanations and designing solutions in 6–8 builds on K–5 experiences and progresses to include constructing explanations and designing solutions supported by multiple sources of evidence consistent with scientific ideas, principles, and theories.
- Apply scientific principles to design an object, tool, process or system.

*Engaging in Argument From Evidence*

Engaging in argument from evidence in 6–8 builds on K–5 experiences and progresses to constructing a convincing argument that supports or refutes claims for either explanations or solutions about the natural and designed world(s).
- Construct an oral and written argument supported by empirical evidence and scientific reasoning to support or refute an explanation or a model for a phenomenon or a solution to a problem.

### DISCIPLINARY CORE IDEAS

*ESS3.C. Human Impacts on Earth Systems*
- Human activities have significantly altered the biosphere, sometimes damaging or destroying natural habitats and causing the extinction of other species. But changes to Earth's environments can have different impacts (negative and positive) for different living things.
- Typically as human populations and per-capita consumption of natural resources increase, so do the negative impacts on Earth, unless the activities and technologies involved are engineered otherwise.

*ESS3.D. Global Climate Change*
- Human activities, such as the release of greenhouse gases from burning fossil fuels, are major factors in the current rise in Earth's mean surface temperature (global warming). Reducing the level of climate change and reducing human vulnerability to whatever climate changes do occur depend on the understanding of climate science, engineering capabilities, and other kinds of knowledge, such as understanding human behavior and applying that knowledge wisely in decision and activities.

---

*Continued*

**Table 4.8.** (*continued*)

CROSSCUTTING CONCEPTS

*Patterns*
- Graphs, charts, and images can be used to identify patterns in data.

*Cause and Effect*
- Relationships can be classified as causal or correlational, and correlation does not necessarily imply causation.
- Cause and effect relationship may be used to predict phenomena in natural or designed systems.

*Stability and Change*
- Stability might be disturbed either by sudden events or gradual changes that accumulate over time.

*COMMON CORE STATE STANDARDS FOR MATHEMATICS*

MATHEMATICAL PRACTICES
- MP1. Make sense of problems and persevere in solving them.
- MP2. Reason abstractly and quantitatively.
- MP3. Construct viable arguments and critique the reasoning of others.
- MP4. Model with mathematics.
- MP5. Use appropriate tools strategically.
- MP6. Attend to precision.
- MP8. Look for and express regularity in repeated reasoning.

MATHEMATICAL CONTENT
- 6.SP.B.4. Display numerical data in plots on a number line, including dot plots, histograms, and box plots.
- 6.SP.B.5.A. Reporting the number of observations.
- 6.SP.B.5.B. Describing the nature of the attribute under investigation, including how it was measured and its units of measurement.

*COMMON CORE STATE STANDARDS FOR ENGLISH LANGUAGE ARTS*

READING STANDARDS
- RI.6.1. Cite textual evidence to support analysis of what the text says explicitly as well as inferences drawn from the text.
- RI.6.2. Determine a central idea of a text and how it is conveyed through particular details; provide a summary of the text distinct from personal opinions or judgments.

*Continued*

**Table 4.8.** (*continued*)

- RI.6.4. Determine the meaning of words and phrases as they are used in a text, including figurative, connotative, and technical meanings.

- RI.6.7. Integrate information presented in different media or formats (e.g., visually, quantitatively) as well as in words to develop a coherent understanding of a topic or issue.

**WRITING STANDARDS**

- W.6.1. Write arguments to support claims with clear reasons and relevant evidence.

- W.6.2. Write informative/explanatory texts to examine a topic and convey ideas, concepts, and information through the selection, organization, and analysis of relevant content.

- W.6.2A. Introduce a topic, organize ideas, concepts and information, using strategies such as definition, classification, comparison/contrast, and cause/effect; include formatting (e.g., headings), graphics (e.g., charts, tables), and multimedia when useful to aiding comprehension.

- W.6.2B. Develop the topic with relevant facts, definitions, concrete details, quotations, or other information and examples.

- W.6.2D. Use precise language and domain specific vocabulary to inform about or explain the topic.

- W.6.7. Conduct short research projects to answer a question, drawing on several sources and refocusing the inquiry when appropriate.

- W.6.8. Gather relevant information from multiple print and digital sources; assess the credibility of each source; and quote or paraphrase the data and conclusions of others while avoiding plagiarism and providing basic bibliographic information for sources.

**SPEAKING AND LISTENING STANDARDS**

- SL.6.1. Engage effectively in a range of collaborative discussions (one-on-one, in groups, and teacher-led) with diverse partners on grade 6 topics, texts, and issues, building on others' ideas and expressing their own clearly.

- SL.6.1.A. Come to discussions prepared, having read or studied required material; explicitly draw on that preparation by referring to evidence on the topic, text, or issue to probe and reflect on ideas under discussion.

- SL.6.1.B. Follow rules for collegial discussions, set specific goals and deadlines, and define individual roles as needed.

- SL.6.2. Interpret information presented in diverse media and formats (e.g., visually, quantitatively, orally) and explain how it contributes to a topic, text, or issue under study.

- SL.6.4. Present claims and findings, sequencing ideas logically and using pertinent descriptions, facts, and details to accentuate main ideas or themes; use appropriate eye contact, adequate volume, and clear pronunciation.

- SL.6.5. Include multimedia components (e.g., graphics, images, music, sound) and visual displays in presentations to clarity information.

*Continued*

**Table 4.8.** (*continued*)

| FRAMEWORK FOR 21ST CENTURY LEARNING |
|---|
| • Interdisciplinary Themes: Global Awareness, Environmental Literacy |
| • Learning and Innovation Skills: Creativity and Innovation, Critical Thinking and Problem Solving, Communication, Collaboration |
| • Information, Media, and Technology Skills: Information Literacy |
| • Life and Career Skills: Initiative and Self-Direction, Productivity and Accountability, Leadership and Responsibility |

**Table 4.9.** Key Vocabulary for Lesson 3

| Key Vocabulary | Definition |
|---|---|
| carbon footprint | the total amount of greenhouse gas emissions, especially carbon dioxide, caused by human activities as a result of the consumption of fossil fuels |
| constraint | something that limits or controls a person's choices |
| emission | something discharged into the air, such as gases from a factory or automobile exhaust |
| fossil fuel | a fuel, such as coal, oil, or natural gas, that was formed in the earth over millions of years from the remains of animals or plants |
| mitigation | the act of lessening the harmful effects of something |
| resource | source of assistance or support; can include printed materials, online materials, or people |

## TEACHER BACKGROUND INFORMATION
### Carbon Footprints

It may be difficult for students to think about how they can help with *global* climate change, as this may seem too abstract or overwhelming. Students may feel helpless to do anything to reverse the effects of climate change. However, students should be able to more readily connect to their local context and identify an environmental problem that contributes to their personal carbon footprints. The Think Globally, Act Locally Challenge provides students an opportunity to identify and define a concrete problem.

We use energy every day. We consume energy when we turn on the lights, drive to school, cook dinner, heat our homes, or watch television. Most of the electricity we use in our households and communities comes from the burning of fossil fuels, which are derived from decayed organic matter in Earth's crust. These fossil fuels are all organic,

or carbon-containing, compounds. Because of the presence of carbon, which creates the greenhouse gas carbon dioxide when we use these fuels, we refer to the amount of fossil fuel–related emissions resulting from the activities of an individual or group as a carbon footprint. Each person's carbon footprint is different. A household's carbon footprint varies by many factors, including home size, modes of transportation, and choices of foods.

## Digital Literacy

Teaching Tolerance provides excellent support resources for teaching digital literacy, including a lesson on evaluating reliable resources at *www.tolerance.org/classroom-resources/tolerance-lessons/evaluating-reliable-sources*. Access the entire set of digital literacy resources at *www.tolerance.org/frameworks/digital-literacy*.

## Picture Books

Picture books, often considered material for lower elementary learners, are excellent resources for the middle school classroom as well. Because of their short length, the underlying structures of the texts are more readily accessible, allowing learners to focus on understanding how the texts are constructed. This knowledge provides the foundation for understanding longer texts. In addition, picture books allow for easy exploration of multiple topics, perspectives, and genres in a short time. They are brief enough to fit easily into an instructional period, and they provide a lot of information and something for students to talk about. Finally, because of the wide range of complexity and readability, picture books readily allow for differentiation among learners.

## COMMON MISCONCEPTIONS

Students will have various types of prior knowledge about the concepts introduced in this lesson. Table 4.10 outlines some common misconceptions students may have concerning these concepts. Because of the breadth of students' experiences, it is not possible to anticipate every misconception that students may bring as they approach this lesson. Incorrect or inaccurate prior understanding of concepts can influence student learning in the future, however, so it is important to be alert to misconceptions such as those presented in the table.

**Table 4.10.** Common Misconceptions About the Concepts in Lesson 3

| Topic | Student Misconception | Explanation |
|---|---|---|
| Bees | Bees are the only plant pollinators. | Although bees are effective and important pollinators, many other species of insects and animals also act as pollinators. |
| Carbon footprint | Eliminating the use of plastic bags will reduce carbon emissions by a great amount. | According to a *Los Angeles Times* article in 2019, even if every American immediately stopped using plastic bags, our annual carbon emissions would decline by only 0.02% (*https://www. latimes.com/opinion/story/2019-11-05/ climate-carbon-emissions-consumer-choices*). It is important to note, however, that plastic bag use has many other environmental implications including increases in solid waste and the danger of plastic consumption by animals. |
| | Driving an electric cars results in no carbon footprint. | The carbon footprint of an electric car depends on the energy source being used to charge its battery. If the electricity is supplied by a coal-fired fuel plant, then the car has a carbon footprint about the same as that of a gasoline-fueled car. |

## PREPARATION FOR LESSON 3

Review the Teacher Background Information section (p. 117) and preview the recommended videos in the Learning Components section to ensure that you have the foundational knowledge about climate change and carbon footprints to teach this lesson. Assemble the materials for the lesson and make copies of student handouts. Prepare an enlarged copy of the bee picture in Figure 4.8 to share electronically or on paper with students. Also gather a supply of nonfiction and informational fiction books that will help students expand their thinking about human impacts on climate change and help them address the module challenge. A list of suggested books for additional reading can be found at the end of this chapter (see p. 129). Assemble several examples of advertisements that advocate for control of climate change and several examples of advertisements that promote activities generally believed to contribute to climate change (the use of fossil fuels, for example). These examples may be in print or other media. You should

also be prepared to display these examples to the class for discussion as part of the ELA Connection in the Explanation section of this lesson. Get approval from the principal for potential school-based plans you might suggest to students for their final challenge.

## LEARNING COMPONENTS

### Introductory Activity/Engagement

**Connection to the Challenge:** Begin the lesson by directing students' attention to the driving question for the module and challenge: How can we develop a local response to address an aspect of human impact on global climate change? Hold a brief discussion of how what they learned in Lesson 2 has influenced their thinking regarding climate change. Review the list of students' climate change questions created in Lesson 1 and have students offer answers to questions based on what they have learned. Note students' responses on the pages you created for questions.

**Science Class:** Have the class examine the enlarged copy of Figure 4.8 that you prepared in advance.

**Figure 4.8. Image of Bee on Flower**

*Note:* A full-color version of this figure is available on the book's Extras page at *www.nsta.org/roadmap-humanimpacts.*

Hold a class discussion, asking students the following questions:

- What do you notice about this bee?

- What is the bee doing?

- How might climate change be associated with bees?

Have students view a video about the impact of climate change on bees and plant pollination, such as "Sting of Climate Change" at *https://climate.nasa.gov/climate_resources/41/ video-sting-of-climate-change* or "Wild Science: Bees and Climate Change" at *www.smith sonianmag.com/smart-news/how-climate-change-messing-bees-ability-pollinate-180956523.*

### STEM Research Notebook Prompt

After the video, ask students to respond to the following questions in their STEM Research Notebooks:

- *What are the big ideas about climate change that you learned from the video?*

- *What new ideas do you have about climate change and how it affects our daily lives?*

- *What new questions do you have about climate change?*

- *Where would humans and the environment be without bees?*

**Mathematics Connection:** Not applicable.

**ELA Connection:** Build student background knowledge and expand their understanding and interest by reading aloud the following award-winning nonfiction and informational fiction picture books on bees:

- *Bees: A Honeyed History*, by Piotr Socha

- *The Bee Book*, by Charlotte Milner

- *The Honeybee*, by Kirsten Hall

- *The Hive Detectives: Chronicle of a Honey Bee Catastrophe*, by Loree Griffin Burns

**Social Studies Connection:** Have students learn where bees are located throughout the world by visiting the "Bees Around the World" website at *www.fao.org/docrep/006/y5110e/ y5110e04.htm#TopOfPage.* Hold a class discussion on the native species in each area and introduction of other species.

## Activity/Exploration

**Science Class:** Ask students to read the Entomology Society of America's "Position Statement on Climate Change" at *www.entsoc.org/sites/default/files/files/Science-Policy/2019/ ESA-Position-Statement-Climate-Change.pdf.* Have students study Figure 1, which focuses

on how climate change can affect insects and subsequently people. Ask students to revisit their response to the last STEM Research Notebook question (*Where would humans and the environment be without bees?*) by using a different-colored writing utensil.

In small groups, have students study Figure 2. Then, ask them the following: "You see one insect life cycle in the 'historical' image, but there is more than one life cycle in the 'climate change' image. What is this image showing you is happening to insect life cycles? What sorts of effects can this have on humans?"

Next, have students focus on Figure 3, which depicts the reproductive potential for the blacklegged tick. Ask students: "With the thought that ticks and other insects, such as mosquitoes, can transmit diseases to humans and other organism, what does this figure tell you about tick distribution?"

Finally, have students examine Figure 4, which depicts how climate change can shift seasonal activity. Tell students: "This figure directly connects to the video that we watched earlier on phenology, bees, blooms, and pollination. In your small groups, study Figure 4 and determine the following" (put questions on overhead or give students a handout if you desire):

1.  *In the "historical" portion of the figure, in what month was*

    a.  *the peak bloom season for the larkspur?* _____

    b.  *the peak of adult activity for the western bumble bee?* _____

2.  *In the "climate change" portion of the figure, in what month was*

    a.  *the peak bloom season for the larkspur?* _____

    b.  *the peak of adult activity for the western bumble bee?* _____

Then, hold a class discussion in which students share their groups' responses and explain why these changes occurred.

## STEM Research Notebook Prompt

Have students respond to the following questions in their STEM Research Notebooks: *How is climate change affecting different organisms? Why is this important to humans?*

**Mathematics Connection:** Provide students with the following scenario:

*Bees are a very important part of our ecosystem. They collect pollen from one flower and spread it to others. Many plants, including fruits and vegetables, can't grow without cross-pollination by bees. Thus, bees are an essential component of our food production worldwide.*

Group students in teams of three or four. Then, instruct them to make a list in their STEM Research Notebooks of all the foods their team members eat regularly. Ask them to discuss as a group whether bees might influence each food's production. Have students circle which foods they think are influenced by bees. Instruct them to calculate the percentage of foods that their group members eat on a regular basis that rely on the work that bees do.

## STEM Research Notebook Prompt

Ask students to respond to the following questions in their STEM Research Notebooks:

- *What does this representation tell you about the impact of bees on your own personal food consumption?*

- *What impact would bees' demise have on your food consumption?*

- *What would happen to the world population as a result of the bees dying?*

- *How would having fewer bees affect the lives of humans?*

**ELA Connection:** Teach students to find and evaluate digital resources for reliability using the digital literacy resources provided in the Teacher Background section or other resources you have identified.

**Social Studies Connection:** Have student teams research the issues that the death of the bees would cause socially and economically worldwide, as well as how they can play a part in saving the bees. Have each team create a plan on how they can save the bees from extinction. *Note:* It is important for students to understand that there are a variety of reasons that bees are dying. Researchers have found that most bees are dying because of *Varroa* mites and tracheal mites. However, climate change also plays a part in their reduction in numbers.

## Explanation

**Science Class:** Ask students to individually develop a visual presentation (e.g., booklet, pamphlet, commercial, public service announcement) to explain the causes and effects of climate change. Students should synthesize data from their STEM Research Notebooks in developing their presentations. Encourage students to include graphs, pictures, figures, data, and descriptions.

After students have created their presentations, divide the class into two or three groups. Have one group set up their team members' visual presentations around the classroom. Then, tell the other two groups to move around the classroom, visiting each presentation. Encourage students to ask questions of the presenters and then complete

a feedback form (see p. 134). Repeat the procedure until all teams have had a turn displaying their presentations. This presentation technique simulates an academic poster session.

**Mathematics Connection:** Have students work in groups of three to compare their findings on the percentage of their food consumption that relies on bees. After students have discussed the factors they considered in arriving at these percentages, have each group calculate an average of the individual group members' percentages. Next, have each group share its average percentage and the range of percentages provided by individuals within the group. Record this information on a class chart. As a class, discuss how averages can mask large variations within groups and how information about the range of items that contribute to the average can be useful. Next, discuss other measures of central tendency, such as median and mode. Have each group calculate the median value for the group, and add this to the class chart.

**ELA Connection:** Reinforce the concept that different genres serve different purposes, with authors choosing genres that best accomplish their purposes. Explore with students various forms of advertising texts, discussing how the different forms serve different purposes, such as reaching and appealing to different audiences. Review with the class techniques used in persuasive writing, including appealing to emotion, using logic, and creating a credible argument. Show students advertising examples from groups advocating for control of climate change and from groups that promote the use of fossil fuels (see Preparation for Lesson 3, p. 119). As a class, analyze the techniques used in the texts, scripts, or visual images of each and whether and how each of these appeals to emotion, uses logic, and creates credible arguments.

**Social Studies Connection:** Have student teams share their research with the class about issues that the death of the bees would cause socially and economically worldwide, along with their plans for saving the bees. Additionally, students can explore the interdependence and interconnectivity of people, places, and environments, including how people around the world adapt to and modify their physical environments. Examples of such adaptation and modification might include types of structures, customs, and activity patterns. Hold a class discussion on how changes in climate might change human activity today.

## Elaboration/Application of Knowledge

**Science Class:** Tell students that sometimes learning about climate change can sound discouraging and overwhelming. However, everyone can do his or her part to help reduce carbon emissions. For their next project, they will use an engineering design process (EDP) to identify a problem that is contributing to carbon dioxide emissions and to design a solution for that problem. They will create a plan to reduce the carbon footprint

of their household, school, community, or state. This can be a project that students can implement in reality, or it can be a conceptualized project. You can modify this assignment according to whether you wish to have students focus on their own individual carbon footprint or address an issue in the community.

Give students the Think Globally, Act Locally Planning Guide (p. 131). As a class, review the guide and the module's EDP, focusing on how each of the steps of the EDP will be used to build the teams' solutions. Students may work individually or in small groups of two or three. Over the course of the next week and a half, students will work on identifying a climate concern and developing a plan to reduce the carbon footprint. Help students define a problem by providing them with examples such as the following:

- Personal

  - Monitor how much waste is left after you eat your lunch. Develop a plan to reduce the amount of waste from your lunch.

  - Think about how you could reduce the carbon impact of the clothes you wear. Develop a plan to recycle your clothing.

- School-based (may need principal's permission to follow through with these ideas)

  - Monitor the amount of waste in the school cafeteria. Develop a plan for reducing waste or composting, or both.

  - Monitor the number of cars that are idling at the school pickup and drop-off area. Develop a plan to reduce the number of idling cars.

  - Monitor the amount of rain runoff. Develop a plan to minimize runoff by developing a rain garden.

Give students time to brainstorm ideas in their STEM Research Notebooks. Tell students they must have their ideas approved by you before starting to develop their plans. For school-based plans, make sure you have obtained approval from the principal before giving students the go-ahead. Over the next several days, students will follow the steps of the module's EDP as listed on the Think Globally, Act Locally Planning Guide:

1. First, students define a problem and research this problem to determine the associated effects on the climate. For instance, if students want to reduce the carbon footprint of the lunches they pack, they could research the following: How many miles did the different foods travel? Are they organic or were pesticides used? How are the lunches packed? Are reusable containers being used?

2. Then, students brainstorm solutions and pick the best one.

3. Students determine any constraints to implementing the solution.

4. Students determine the available resources, including experts they can interview who might be able to help them develop a solution.

5. Next, students develop their plans to mitigate their defined problems. Students should draw diagrams and make lists of required materials. The plan needs to include how they will monitor the defined problem and how they will collect data to determine whether the plan is working. *Note:* The monitoring and data collection plans can change as students develop their ideas.

6. Now, students test their plans or begin monitoring. Give students time to discuss their plans with classmates so that they can receive feedback.

7. Students decide how they can improve their plans once they determine what is and is not working.

8. Students revise their plans.

9. Students present their final projects. They could present them to the class, or you could hold a special showcase and invite members of the community and parents to see the presentations. Use the Think Globally, Act Locally Presentation Rubric (pp. 132–133) to grade the final presentations.

Finally, have students write up their plans, explaining the defined problem, their solution to the problem, and why they think it is important to solve this problem.

**Mathematics Connection:** Remind students of the definition of *carbon footprint*. Ask, "Do you know what your carbon footprint is?" To help students think about the impact their family has on their environment and connect to the challenge, have them calculate their carbon footprints. One example of a student-friendly carbon emissions calculator can be found at *www.ei.lehigh.edu/learners/cc/carboncalc.html*. See Figure 4.9 for an example from this website. Another calculator can be found at *www.meetthegreens.org/features/carbon-calculator.html*.

**Figure 4.9.** Sample Analysis Output for Carbon Footprint

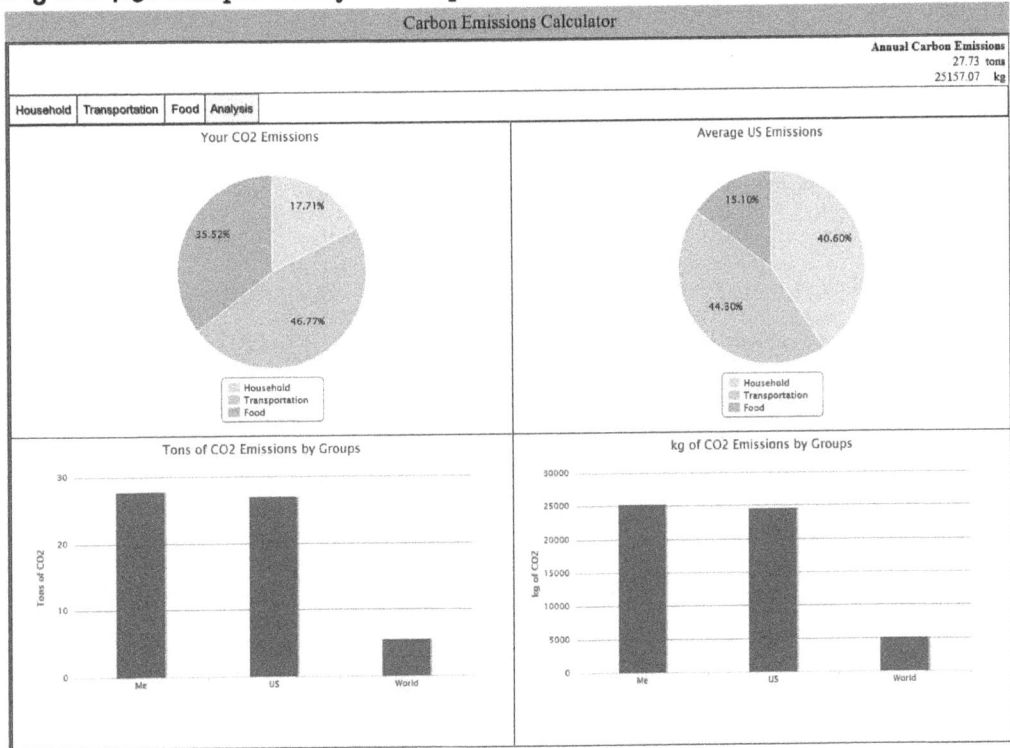

*Source: www.ei.lehigh.edu/learners/cc/carboncalc.html.*

*Note:* A full-color version of this figure is available on the book's Extras page at *www.nsta.org/roadmap-humanimpacts.*

After students calculate their carbon footprints, hold a discussion with students about things they are doing in their daily lives that they could change to help reduce their own personal carbon emissions. Show students a video about how to reduce their carbon footprint, such as "What Is a Carbon Footprint? What Can You Do About Yours?" at *www.youtube.com/watch?v=YseZXKfT_yY* or "What Are Carbon Footprints" at *www.youtube.com/watch?v=Dwkh46MZuIc.*

**ELA Connection:** Read aloud an accessible book such as *Energy Island: How One Community Harnessed the Wind and Changed Their World,* by Allan Drummond, to introduce the idea of social activism in response to climate change. Have students conduct internet research about various social and political actions around the world aimed at reducing carbon emissions. Then, hold a class discussion on what they found.

**Social Studies Connection:** Have students take the post-test and draw a new picture of their environment based on what they have learned in this module. They should then compare their answers with their pre-test responses. Have another gallery walk for the new drawings. Ask students how their ideas about the environment and humans' role in the environment have changed since the beginning of this unit. You might have students write a final reflection on the negative and positive impacts that humans can have on climate change.

## Evaluation/Assessment

Students may be assessed on the following performance tasks and other measures listed.

*Performance Tasks*

- Think Globally, Act Locally Presentation Rubric (pp. 132–133)

- Visual Presentation Peer Feedback Form (p. 134)

- Climate Change Post-Test (p. 135)

- Environment drawing (p. 137)

*Other Measures*

- STEM Research Notebook Entry Rubric (see Lesson 1, p. 72)

- Participation in class discussions

## INTERNET RESOURCES

Evaluating reliable resources
- *www.tolerance.org/classroom-resources/tolerance-lessons/evaluating-reliable-sources*

Digital literacy
- *www.tolerance.org/frameworks/digital-literacy*

Opinion: Advice for the climate-conscious consumer: Don't sweat the small stuff
- *https://www.latimes.com/opinion/story/2019-11-05/climate-carbon-emissions-consumer-choices*

"Sting of Climate Change" video
- *https://climate.nasa.gov/climate_resources/41/video-sting-of-climate-change*

"Wild Science: Bees and Climate Change"
- *www.smithsonianmag.com/smart-news/how-climate-change-messing-bees-ability-pollinate-180956523*

"Bees Around the World"
- *www.fao.org/docrep/006/y5110e/y5110e04.htm#TopOfPage*

Entomology Society for America's position statement on climate change
- *www.entsoc.org/sites/default/files/files/Science-Policy/2019/ESA-Position-Statement-Climate-Change.pdf*

Carbon emissions calculators
- *www.ei.lehigh.edu/learners/cc/carboncalc.html*

- *www.meetthegreens.org/features/carbon-calculator.html*

"What Is a Carbon Footprint? What Can You Do About Yours?" video
- *www.youtube.com/watch?v=YseZXKfT_yY*

"What Are Carbon Footprints" video
- *www.youtube.com/watch?v=Dwkh46MZuIc*

## SUGGESTED BOOKS

- *Analyzing Climate Change: Asking Questions, Evaluating Evidence, and Designing Solutions*, by Philip Steele (Cavendish Square, 2018)

- *Bees: A Honeyed History*, by Piotr Socha (Harry N. Abrams, 2017)

- *Energy Island: How One Community Harnessed the Wind and Changed Their World*, by Allan Drummond (Farrar, Straus and Giroux, 2011)

- *Exodus*, by Julie Bertagna (Pan Macmillan, 2002/2017)

- *How We Know What We Know About Our Changing Climate*, by Lynne Cherry and Gary Braasch (Dawn Publications, 2008/2010)

- *It's Getting Hot in Here: The Past, Present, and Future of Climate Change*, by Bridget Heos (Houghton Mifflin Harcourt, 2016)

- *The Bee Book*, by Charlotte Milner (DK Children, 2018)

- *The Hive Detectives: Chronicle of a Honey Bee Catastrophe*, by Loree Griffin Burns (Scholastic, 2014)

- *The Honeybee*, by Kirsten Hall (Atheneum Books for Young Readers, 2018)

- *The Last Wild*, by Piers Torday (Puffin Books, 2014/2015)

- *The Magic School Bus and the Climate Challenge*, by Joanna Cole (Scholastic Press, 2010)

- *What Is Climate Change?*, by Gail Herman (Penguin Random House, 2018)

## REFERENCES

U.N. Intergovernmental Panel on Climate Change (IPCC). 2014. Summary for policymakers. In *Climate Change 2014: Mitigation of Climate Change*. Contribution of Working Group III to the Fifth Assessment Report of the Intergovernmental Panel on Climate Change, ed. O. Edehofer et al., Cambridge University Press, Cambridge, UK.

Moseley, C., B. Perrotta, and J. Utley. 2010. The Draw-An-Environment Test Rubric (DAET-R): Exploring preservice teachers' mental models of the environment. *Environmental Education Research* 16 (2): 189–208. *http://dx.doi.org/10.1080/13504620903548674.*

National Research Council (NRC). 2007. *Taking science to school: Learning and teaching science in grades K–8.* Washington, DC: National Academies Press.

National Research Council (NRC). 2012. *A framework for K–12 science education: Practices, crosscutting concepts, and core ideas.* Washington, DC: National Academies Press.

NGSS Lead States. 2013. *Next Generation Science Standards: For states, by states.* Washington, DC: National Academies Press. *www.nextgenscience.org/next-generation-sciencestandards.*

U.S. Global Change Research Program (USGCRP). 2009. *Climate literacy: The essential principles of climate science.* Washington, DC: USGCRP. *https://downloads.globalchange.gov/Literacy/climate_literacy_highres_english.pdf.*

Name: _____        Date: _____

# THINK GLOBALLY, ACT LOCALLY PLANNING GUIDE

**Purpose:** For this project, your goal is to identify a problem contributing to global climate change. This can be a BIG problem or a smaller problem. This could be something that you or someone else can do to reduce your personal carbon footprint on the planet. Or, it could be a policy for your school, neighborhood, community, or state to do as a group.

**What to do:** This activity has TWO PARTS.

**Part A.** Complete the following Think Globally, Act Locally Challenge action items, recording your notes in your STEM Research Notebook:

| Action Item | Due Date |
|---|---|
| 1. **Define:** Define your problem. | |
| 2. **Learn:** Brainstorm solutions and pick your best one! | |
| 3. **Plan:** Determine your constraints. | |
| 4. **Plan:** Determine your resources. Which experts can you ask for help to develop your solution? | |
| 5. **Try:** Develop a plan to mitigate the problem, including a labeled sketch, a list of needed materials, how you will monitor the problem, and how you will collect data. | |
| 6. **Test:** Test your plan or begin monitoring the problem. | |
| 7. **Decide:** Decide how you can improve your plan. | |
| 8. **Decide:** Revise your plan. | |
| 9. Put your plan into action. | |
| 10. Present your plan. | |

**Part B.** Write up your plan in your STEM Research Notebook, explaining the following:
- Your identified problem
- Your solution to the problem
- Why you think it is important to solve this problem

## Think Globally, Act Locally Challenge Rubric

Name: _____

| Criteria | Beginning/Does Not Meet Expectations (1 point) | Meets Expectations (2 points) | Advanced (3 points) | Score |
|---|---|---|---|---|
| HUMAN IMPACT | You identified the problem but the focus on human impacts is not clear. | You identified the problem with a focus on human impacts. | You identified the problem with a focus on human impacts; problem is tied to research. | |
| DESIGN SOLUTION | Your design solution includes a plan. You did not seek advice from experts. | Your design solution includes a well-thought-out plan. You sought advice from experts. | Your design solution includes a well-thought-out plan. You sought advice from experts. Your solution is based on evidence and research. | |
| TEST AND IMPROVE | You did not test your design, receive feedback from others, or improve your design. | You tested your design, received feedback from others, and improved your design. | You tested your design thoroughly and creatively, received feedback from others, and improved your design in creative ways that addressed specific feedback you received. | |
| MONITOR | You have an incomplete plan to monitor the problem and collect data to monitor the solution. | Your plan included a way to monitor the defined problem and how to collect data to monitor your solution. | Your plan included a way to monitor the defined problem and how to collect data to monitor your solution. You went beyond the plan to monitor and actually collected some data. | |
| PRESENTATION | Presentation is not very clear in some areas. | Presentation is accurate and presented so that audience can understand. | Presentation is original, appealing, accurate, well thought out, and clearly presented so that the audience can understand. | |

*Continued*

Name: _____

## Think Globally, Act Locally Challenge Rubric (*continued*)

| Criteria | Beginning/Does Not Meet Expectations (1 point) | Meets Expectations (2 points) | Advanced (3 points) | Score |
|---|---|---|---|---|
| MANUSCRIPT | Your manuscript may be missing your identified problem, your solution to the problem, or why you think it is important. It is not supported with references. | Your manuscript includes your identified problem, your solution to the problem, and why you think it is important. It is supported with some references. | Your manuscript includes your identified problem, your solution to the problem, and why you think it is important. It is well supported with references. | |
| PROFESSIONALISM | Presentation and manuscript are presented with most components; either may contain multiple grammatical and spelling errors. | Presentation and manuscript are presented with all components and contain minor grammatical and spelling errors. | Presentation and manuscript are presented with all components and are free of grammatical and spelling errors. | |

TOTAL SCORE: _____

COMMENTS:

**STUDENT HANDOUT**

## VISUAL PRESENTATION PEER FEEDBACK FORM

Name of Presenter: _____    My Name: _____

Problem Clearly Stated:

Solution Clearly Stated:

Clarity:

Comments:

*Note:* 1 Earth = element is absent; 2 Earths = needs improvement; 3 Earths = good work; 4 Earths = great work and exceeds expectations!

Name: _____     Date: _____

# CLIMATE CHANGE POST-TEST

1. What is climate?

2. What is weather?

3. What do you know about climate change?

4. How well do you feel you understand the issue of climate change? **(circle one)**

   Very well        Fairly well        Not very well        Not at all        No opinion

5. How well do you feel you understand the issue of global warming? **(circle one)**

   Very well        Fairly well        Not very well        Not at all        No opinion

Name: _____     Date: _____

# CLIMATE CHANGE POST-TEST

6. Do you believe increases in Earth's temperature over the last century are due more to … **(select one)**

    a. Human activities
    b. Natural causes
    c. Don't know

7. In your opinion, when will the effects of climate change begin to happen? **(select one)**

    a. They will never happen.
    b. They will not happen in my lifetime, but future generations will feel their effects.
    c. They will start happening within my lifetime.
    d. They will start happening in the next few years.
    e. They are happening now.

True/False **(circle one)**

8. Weather and climate are the same thing.　　　　　　　　　　　True　or　False

9. Global warming and the greenhouse effect are the same thing.　　True　or　False

10. Climate change and global warming are the same thing.　　　　　True　or　False

11. Carbon dioxide is a greenhouse gas.　　　　　　　　　　　　　True　or　False

12. Greenhouse gases capture heat and warm Earth and its atmosphere.　True　or　False

13. Greenhouse gases are caused only by humans.　　　　　　　　　True　or　False

14. Scientists agree about whether climate change is happening.　　　True　or　False

15. Climate change is happening as a result of an increase in greenhouse gases in the atmosphere.　　　　　　　　　　　　　True　or　False

16. There is nothing that I can do to lessen the effects of climate change.　True　or　False

Name: _____

Date: _____

**STUDENT HANDOUT**

# DRAWING OF MY ENVIRONMENT

In the space below, draw a picture of your environment.

*Source:* Adapted from Moseley, Perrotta, and Utley (2010).

# 5

# TRANSFORMING LEARNING WITH HUMAN IMPACTS ON OUR CLIMATE AND THE *STEM ROAD MAP CURRICULUM SERIES*

*Carla C. Johnson*

This chapter serves as a conclusion to the Human Impacts on Our Climate integrated STEM curriculum module, but it is just the beginning of the transformation of your classroom that is possible through use of the *STEM Road Map Curriculum Series*. In this book, many key resources have been provided to make learning meaningful for your students through integrating science, technology, engineering, and mathematics, as well as social studies and English language arts, into powerful problem- and project-based instruction. First, the Human Impacts on Our Climate curriculum is grounded in the latest theory of learning for students in grade 6 specifically. Second, as your students work through this module, they engage in using an engineering design process (EDP) and build prototypes like engineers and STEM professionals in the real world. Third, students acquire important knowledge and skills grounded in national academic standards in mathematics, English language arts, science, and 21st century skills that will enable their learning to be deeper, retained longer, and applied throughout, illustrating the critical connections within and across disciplines. Finally, authentic formative assessments, including strategies for differentiation and addressing misconceptions, are embedded within the curriculum activities.

The Human Impacts on Our Climate curriculum in the Cause and Effect STEM Road Map theme can be used in single-content classrooms (e.g., science) where there is only one teacher or expanded to include multiple teachers and content areas across classrooms. Through the exploration of the Think Globally, Act Locally Challenge, students

engage in a real-world STEM problem on the first day of instruction and gather necessary knowledge and skills along the way in the context of solving the problem.

The other topics in the *STEM Road Map Curriculum Series* are designed in a similar manner, and NSTA Press has additional volumes in this series for this and other grade levels and plans to publish more. The volumes covering Innovation and Progress have been published and are as follows:

- *Amusement Park of the Future, Grade 6*
- *Construction Materials, Grade 11*
- *Harnessing Solar Energy, Grade 4*
- *Transportation in the Future, Grade 3*
- *Wind Energy, Grade 5*

The volumes covering The Represented World have also been published and are as follows:

- *Car Crashes, Grade 12*
- *Improving Bridge Design, Grade 8*
- *Investigating Environmental Changes, Grade 2*
- *Packaging Design, Grade 6*
- *Patterns and the Plant World, Grade 1*
- *Radioactivity, Grade 11*
- *Rainwater Analysis, Grade 5*
- *Swing Set Makeover, Grade 3*

In addition, several volumes covering Cause and Effect have been published:

- *Influence of Waves, Grade 1*
- *Natural Hazards, Grade 2*
- *Physics in Motion, Grade K*

The tentative list of other books includes the following themes and subjects:

- Cause and Effect (*continued*)
  - Healthy living
  - The changing Earth

- Sustainable Systems

  - Composting: Reduce, reuse, recycle

  - Creating global bonds

  - Hydropower efficiency

  - System interactions

- Optimizing the Human Experience

  - Genetically modified organisms

  - Mineral resources

  - Rebuilding the natural environment

If you are interested in professional development opportunities focused on the STEM Road Map specifically or integrated STEM or STEM programs and schools overall, contact the lead editor of this project, Dr. Carla C. Johnson (*carlacjohnson@ncsu.edu*), professor of science education in the College of Education and Office of Research and Innovation Faculty Research Fellow at North Carolina State University in Raleigh. Someone from the team will be in touch to design a program that will meet your individual, school, or district needs.

# APPENDIX

## CONTENT STANDARDS ADDRESSED IN THIS MODULE

### *NEXT GENERATION SCIENCE STANDARDS*

Table A1 (p. 144) lists the science and engineering practices, disciplinary core ideas, and crosscutting concepts this module addresses. The supported performance expectations are as follows:

- MS-ESS3-3. Apply scientific principles to design a method for monitoring and minimizing a human impact on the environment.

- MS-ESS3-4. Construct an argument supported by evidence for how increases in human population and per-capita consumption of natural resources impact Earth's systems.

- MS-ESS3-5. Ask questions to clarify evidence of the factors that have caused the rise in global temperatures over the past century.

- MS-ETS1. Define the criteria and constraints of a design problem with sufficient precision to ensure a successful solution, taking into account relevant scientific principles and potential impacts on people and the natural environment.

**Table A1.** *Next Generation Science Standards (NGSS)*

| Science and Engineering Practices |
|---|

### ASKING QUESTIONS AND DEFINING PROBLEMS

Asking questions and defining problems in grades 6–8 builds on grades K–5 experiences and progresses to specifying relationships between variables, and clarifying arguments and models.
- Ask questions to identify and clarify evidence of an argument.

### ANALYZING AND INTERPRETING DATA

Analyzing data in 6–8 builds on K–5 and progresses to extending quantitative analysis to investigations, distinguishing between correlation and causation, and basic statistical techniques of data and error analysis.
- Analyze and interpret data to determine similarities and differences in findings.

### CONSTRUCTING EXPLANATIONS AND DESIGNING SOLUTIONS

Constructing explanations and designing solutions in 6–8 builds on K–5 experiences and progresses to include constructing explanations and designing solutions supported by multiple sources of evidence consistent with scientific ideas, principles, and theories.
- Apply scientific principles to design an object, tool, process or system.

### ENGAGING IN ARGUMENT FROM EVIDENCE

Engaging in argument from evidence in 6–8 builds on K–5 experiences and progresses to constructing a convincing argument that supports or refutes claims for either explanations or solutions about the natural and designed world(s).
- Construct an oral and written argument supported by empirical evidence and scientific reasoning to support or refute an explanation or a model for a phenomenon or a solution to a problem.

| Disciplinary Core Ideas |
|---|

### ESS2.D. WEATHER AND CLIMATE

- Because these patterns are so complex, weather can only be predicted probabilistically.

### ESS3.C. HUMAN IMPACTS ON EARTH SYSTEMS

- Human activities have significantly altered the biosphere, sometimes damaging or destroying natural habitats and causing the extinction of other species. But changes to Earth's environments can have different impacts (negative and positive) for different living things.

- Typically as human populations and per-capita consumption of natural resources increase, so do the negative impacts on Earth, unless the activities and technologies involved are engineered otherwise.

### ESS3.D. GLOBAL CLIMATE CHANGE

- Human activities, such as the release of greenhouse gases from burning fossil fuels, are major factors in the current rise in Earth's mean surface temperature (global warming). Reducing the level of climate change and reducing human vulnerability to whatever climate changes do occur depend on the understanding of climate science, engineering capabilities, and other kinds of knowledge, such as understanding human behavior and applying that knowledge wisely in decision and activities.

*Continued*

**Table A1.** (*continued*)

| Crosscutting Concepts |
|---|

**PATTERNS**

- Graphs, charts, and images can be used to identify patterns in data.

**CAUSE AND EFFECT**

- Relationships can be classified as causal or correlational, and correlation does not necessarily imply causation.

- Cause and effect relationship may be used to predict phenomena in natural or designed systems.

**STABILITY AND CHANGE**

- Stability might be disturbed either by sudden events or gradual changes that accumulate over time.

*Source:* NGSS Lead States. 2013. *Next Generation Science Standards: For states, by states.* Washington, DC: National Academies Press. *www.nextgenscience.org/next-generation-science-standards.*

## Table A2. Common Core Mathematics and English Language Arts (ELA) Standards

### MATHEMATICAL PRACTICES

- MP1. Make sense of problems and persevere in solving them.
- MP2. Reason abstractly and quantitatively.
- MP3. Construct viable arguments and critique the reasoning of others.
- MP4. Model with mathematics.
- MP5. Use appropriate tools strategically.
- MP6. Attend to precision.
- MP8. Look for and express regularity in repeated reasoning.

### MATHEMATICAL CONTENT

- 6.SP.A.3. Recognize that a measure of center for a numerical data set summarizes all of its values with a single number, while a measure of variation describes how its values vary with a single number.
- 6.SP.B.4. Display numerical data in plots on a number line, including dot plots, histograms, and box plots.
- 6.SP.5. Summarize numerical data sets in relation to their context.
- 6.SP.B.5.A. Reporting the number of observations.
- 6.SP.B.5.B. Describing the nature of the attribute under investigation, including how it was measured and its units of measurement.

### READING STANDARDS

- RI.6.1. Cite textual evidence to support analysis of what the text says explicitly as well as inferences drawn from the text.
- RI.6.2. Determine a central idea of a text and how it is conveyed through particular details; provide a summary of the text distinct from personal opinions or judgments.
- RI.6.4. Determine the meaning of words and phrases as they are used in a text, including figurative, connotative, and technical meanings.
- RI.6.7. Integrate information presented in different media or formats (e.g., visually, quantitatively) as well as in words to develop a coherent understanding of a topic or issue.

### WRITING STANDARDS

- W.6.1. Write arguments to support claims with clear reasons and relevant evidence.
- W.6.2. Write informative/explanatory texts to examine a topic and convey ideas, concepts, and information through the selection, organization, and analysis of relevant content.
- W.6.2A. Introduce a topic, organize ideas, concepts and information, using strategies such as definition, classification, comparison/contrast, and cause/effect; include formatting (e.g., headings), graphics (e.g., charts, tables), and multimedia when useful to aiding comprehension.
- W.6.2B. Develop the topic with relevant facts, definitions, concrete details, quotations, or other information and examples.
- W.6.2D. Use precise language and domain specific vocabulary to inform about or explain the topic.
- W.6.2F. Provide a concluding statement or section that follows from the information or explanation presented.
- W.6.7. Conduct short research projects to answer a question, drawing on several sources and refocusing the inquiry when appropriate.
- W.6.8. Gather relevant information from multiple print and digital sources; assess the credibility of each source; and quote or paraphrase the data and conclusions of others while avoiding plagiarism and providing basic bibliographic information for sources
- WHST.6 0.2.D. Use precise language and domain-specific vocabulary to inform about or explain the topic.
- WHST.6-8.8. Gather relevant information from multiple print and digital sources, using search terms effectively; assess the credibility and accuracy of each source; and quote or paraphrase the data and conclusions of others while avoiding plagiarism and following a standard format for citation

*Continued*

**Table A2.** (*continued*)

| | SPEAKING AND LISTENING STANDARDS |
|---|---|
| | • SL.6.1. Engage effectively in a range of collaborative discussions (one-on-one, in groups, and teacher-led) with diverse partners on grade 6 topics, texts, and issues, building on others' ideas and expressing their own clearly. |
| | • SL.6.1.A. Come to discussions prepared, having read or studied required material; explicitly draw on that preparation by referring to evidence on the topic, text, or issue to probe and reflect on ideas under discussion. |
| | • SL.6.1.B. Follow rules for collegial discussions, set specific goals and deadlines, and define individual roles as needed. |
| | • SL.6.1.C. Pose and respond to specific questions with elaboration and detail by making comments that contribute to the topic, text, or issue under discussion. |
| | • SL.6.2. Interpret information presented in diverse media and formats (e.g., visually, quantitatively, orally) and explain how it contributes to a topic, text, or issue under study. |
| | • SL.6.4. Present claims and findings, sequencing ideas logically and using pertinent descriptions, facts, and details to accentuate main ideas or themes; use appropriate eye contact, adequate volume, and clear pronunciation. |
| | • SL.6.5. Include multimedia components (e.g., graphics, images, music, sound) and visual displays in presentations to clarify information. |

*Source:* National Governors Association Center for Best Practices and Council of Chief State School Officers (NGAC and CCSSO). 2010. *Common core state standards.* Washington, DC: NGAC and CCSSO.

**Table A3.** 21st Century Skills From the Framework for 21st Century Learning

| 21st Century Skills | Learning Skills and Technology Tools | Teaching Strategies | Evidence of Success |
|---|---|---|---|
| **INTERDISCIPLINARY THEMES** | • Global Awareness<br>• Environmental Literacy | • Discuss differences between weather and climate.<br>• Identify how humans can influence climate change. | • Students created multimedia presentations on their global warming findings.<br>• Students completed STEM Research Notebook entries. |
| **LEARNING AND INNOVATION SKILLS** | • Creativity and Innovation<br>• Critical Thinking and Problem Solving<br>• Communication<br>• Collaboration | • Help students analyze data and interpret trends in average global temperatures and develop scientific arguments.<br>• Have students analyze ways to reduce their carbon footprints. | • Students developed scientific arguments to present evidence related to climate change indicators.<br>• Students developed mitigation plans. |
| **INFORMATION, MEDIA, AND TECHNOLOGY SKILLS** | • Information Literacy | • Ask students to apply research skills to locate historical and current global temperature data.<br>• Provide opportunities for students to present their mitigation plans. | • Students gave presentations. |
| **LIFE AND CAREER SKILLS** | • Initiative and Self-Direction<br>• Productivity and Accountability<br>• Leadership and Responsibility | • Provide guidelines for effective peer critique and how students can use this feedback to improve their presentations.<br>• Establish collaborative learning expectations.<br>• Scaffold completion of tasks. | • Students completed a mitigation plan on time and with evidence of collaboration by the whole group.<br>• Students reflected on presentation of mitigation plan in their STEM Research Notebooks. |

*Source:* Partnership for 21st Century Learning. Battelle for Kids. 2015. Framework for 21st Century Learning. *www. battelleforkids.org/networks/p21/frameworks-resources.*

**Table A4.** English Language Development (ELD) Standards

---

**ELD STANDARD 1: SOCIAL AND INSTRUCTIONAL LANGUAGE**

English language learners communicate for Social and Instructional purposes within the school setting.

**ELD STANDARD 2: THE LANGUAGE OF LANGUAGE ARTS**

English language learners communicate information, ideas and concepts necessary for academic success in the content area of Language Arts.

**ELD STANDARD 3: THE LANGUAGE OF MATHEMATICS**

English language learners communicate information, ideas and concepts necessary for academic success in the content area of Mathematics.

**ELD STANDARD 4: THE LANGUAGE OF SCIENCE**

English language learners communicate information, ideas and concepts necessary for academic success in the content area of Science.

**ELD STANDARD 5: THE LANGUAGE OF SOCIAL STUDIES**

English language learners communicate information, ideas and concepts necessary for academic success in the content area of Social Studies.

---

*Source:* WIDA. 2012. 2012 amplification of the English language development standards: Kindergarten–grade 12. *https://wida.wisc.edu/teach/standards/eld.*

# INDEX

Page numbers printed in **boldface type** indicate tables, figures, or handouts.